Monu...
Kakega...
Shizuok...
Prefectu...

JN119572

Fujikawa Bridge
Fuji City,Shizuoka Prefecture

Triangle and Quadrilateral

Satsuma Kiriko Tower
Kagoshima City,
Kagoshima Prefecture

Aomori Prefecture Tourist Center
Aomori City, Aomori Prefecture

1

Table of Contents

Let's learn mathematics together.

Daiki

Yui

Nanami

Hiroto

① Thinking Competency

Competency to think the same or similar way
Competency to relate what you are learning to what you have learned before and think in the same or similar way

Competency to find rules
Competency to analyze various numbers and expressions and investigate any rules

Competency to explain the reason
Competency to explain the reason why, based on learned rules and important ideas

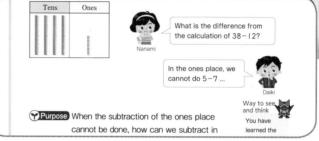

Tens	Ones

What is the difference from the calculation of 38−12?

Nanami

In the ones place, we cannot do 5−7 ...

Daiki

Way to see and think

You have learned the

Purpose When the subtraction of the ones place cannot be done, how can we subtract in

② Judgement Competency

Competency to find mistakes
Competency to find the over generalization of conventions and rules

Competency to find the correct
Competency to find the correct way by comparing the numbers or the quantities

Competency to compare ideas and way of thinking
Competency to find the same or different ways of thinking by comparing friends' ideas and your own ideas

Want to discuss

 Find the mistakes in the following processes. Let's talk about how to calculate correctly.

① 608 − 3   ② 524 − 17

③ Representation Competency

Competency to represent sentences with pictures or expressions
Competency to read problem sentences, draw pictures, and represent them using expressions

Competency to represent pictures or expressions with sentences
Competency to understand pictures or expressions and represent problem sentences

Competency to communicate with your friends
Competency to communicate your ideas to friends in simpler and easier manners

Want to represent

① Yui has the following idea. Let's continue writing her idea.

 Yui's idea

If I put two cups of B into cup C, then there will

Let's find monsters.

Monsters
which represent ways of thinking in mathematics

Setting the unit.

Once you have decided one unit, you can represent how many.

Unit

If you try to separate...

Decomposing numbers by place value and dividing figures make it easier to think about problems.

Separate

If you represent in other way...

If you represent in other expression, diagram, table, etc., it is easier to understand.

Other way

Can you do the same or similar way?

If you find something the same or similar, then you can understand.

Looks same

You wonder why?

Why this happens? If you communicate the reasons in order, it will be easier to understand for others.

Why

If you try to arrange...

You can compare if you align the number place and align the unit.

Align

If you try to summarize...

It is easier to understand if you put the numbers in groups of 10 or summarize in a table or graph.

Summarize

If you try to change the number or figure...

If you try to change the problem a little, you can better understand the problem or find a new problem.

Change

Is there a rule?

If you examine a few examples, then you can find out whether there is a rule.

Rule

Ways to think learned in the 1st grade

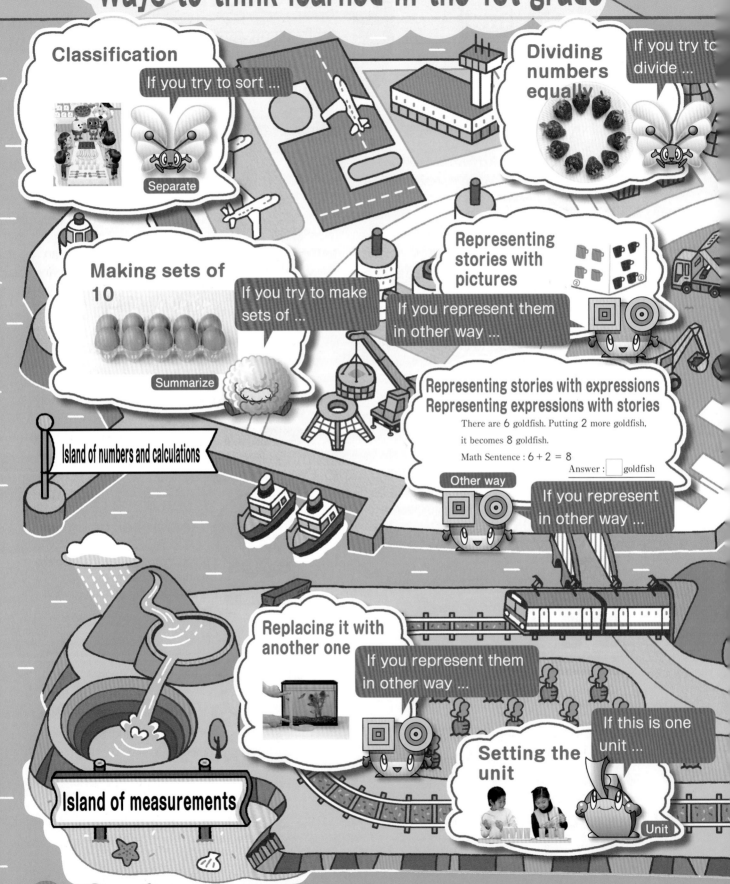

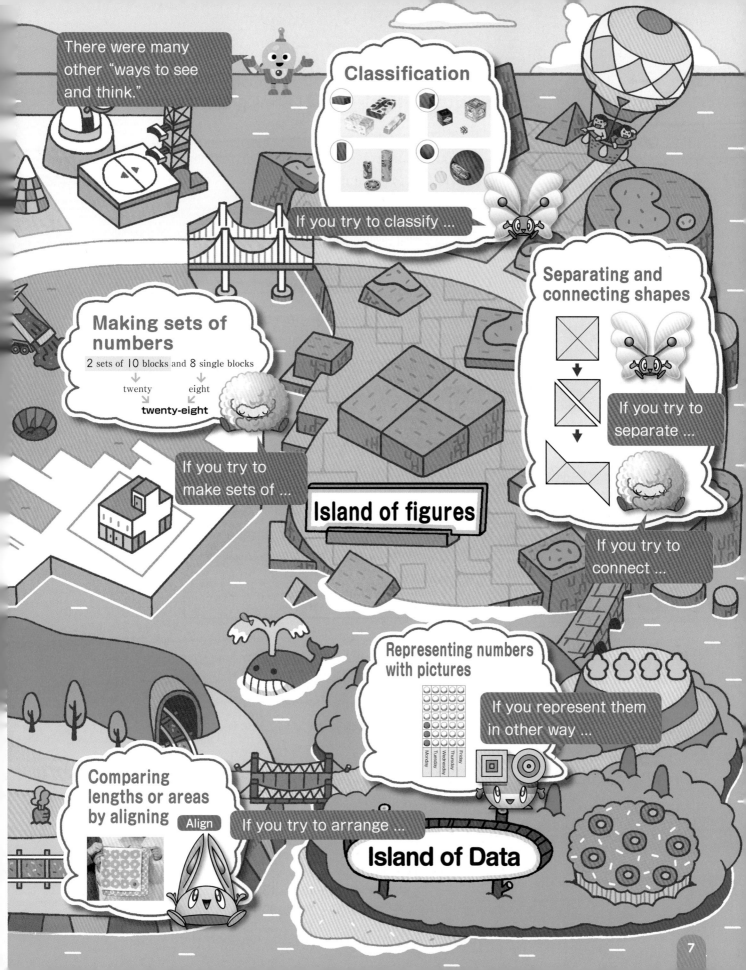

1 Self directed learning: Learning on your own initiative.

Find the ? Problem

If you find a problem in your life or mathematics, you will like to solve it.

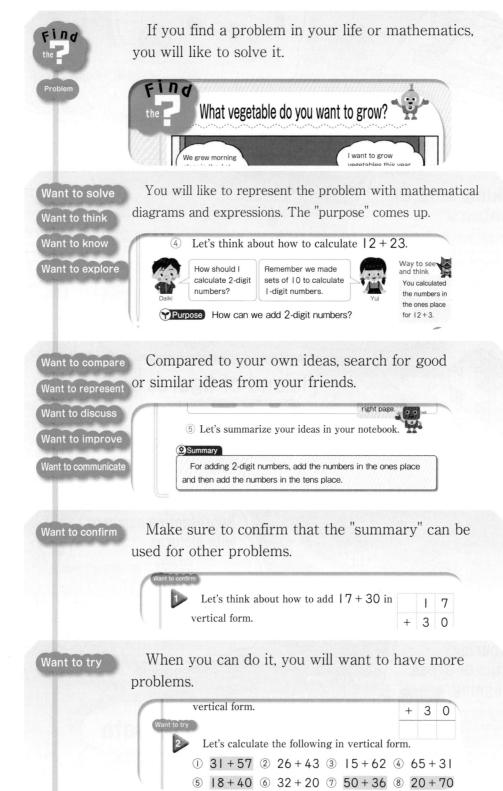

Find the ? What vegetable do you want to grow?

We grew morning ... | I want to grow vegetables this year ...

Want to solve
Want to think
Want to know
Want to explore

You will like to represent the problem with mathematical diagrams and expressions. The "purpose" comes up.

④ Let's think about how to calculate 12 + 23.

Daiki: How should I calculate 2-digit numbers? | Remember we made sets of 10 to calculate 1-digit numbers. | Yui

Way to see and think
You calculated the numbers in the ones place for 12 + 3.

Purpose How can we add 2-digit numbers?

Want to compare
Want to represent
Want to discuss
Want to improve
Want to communicate

Compared to your own ideas, search for good or similar ideas from your friends.

right page.

⑤ Let's summarize your ideas in your notebook.

Summary
For adding 2-digit numbers, add the numbers in the ones place and then add the numbers in the tens place.

Want to confirm

Make sure to confirm that the "summary" can be used for other problems.

Want to confirm

1 Let's think about how to add 17 + 30 in vertical form.

```
  1 7
+ 3 0
```

Want to try

When you can do it, you will want to have more problems.

vertical form.

```
+ 3 0
```

Want to try

2 Let's calculate the following in vertical form.
① 31 + 57 ② 26 + 43 ③ 15 + 62 ④ 65 + 31
⑤ 18 + 40 ⑥ 32 + 20 ⑦ 50 + 36 ⑧ 20 + 70

⭐2 Dialogue learning: Learning together with friends.

As learning progresses, you will want to know what your friends are thinking. Also, you will like to share your own ideas with your friends. Let's discuss with the person next to you, in groups, or with the whole class.

Want to discuss

We have learned how to tell time in the 1st grade.

How do the long hand and the short hand move?

Want to explain

How about aligning their edges?

But we have to move them.

Can you compare them as you have learned in the 1st grade?

We compared by counting the number of some units in the 1st grade.

Then we can compare the lengths without moving them.

⭐3 Deep learning: Usefulness and efficiency of what you learned.

Let's cherish the feeling "I want to know more." and "Can I do this in another case?" Let's deepen learning in life and mathematics.

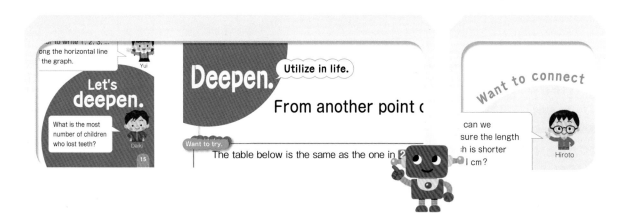

...r to write 1, 2, 3, ...
...ong the horizontal line
...the graph.

Yui

Let's deepen.

What is the most number of children who lost teeth?

Daiki

15

Deepen. Utilize in life.

From another point (

Want to try.

The table below is the same as the one in

Want to connect

...can we
...sure the length
...h is shorter
...l cm?

Hiroto

Symbols in this textbook

④ Let's think about how to calculate $12 + 23$.

Daiki: How should I calculate 2-digit numbers?

Remember we made sets of 10 to calculate 1-digit numbers.

Yui

Way to see and think
You calculated the numbers in the ones place for $12 + 3$.

Purpose How can we add 2-digit numbers?

Summary

For adding 2-digit numbers, add the numbers in the ones place and then add the numbers in the tens place.

Purpose

When you think about problems, you sometimes ask **"Why?"**
That is the **purpose** you should learn.

Summary

The rules you could find through learning the new content are **summarized**.

The position of 2 in 235 is called the **hundreds place**.

Doctor

Important words and rules are taught by Doctor.

What you can do now

Usefulness and efficiency of learning

What you can do now
Usefulness and efficiency of learning

After learning in each chapter, you can check **"what you can do now"** and solve the problems with **"usefulness and efficiency of learning"** in each page.

Reflect · Connect

You can **reflect** what you have learned and **connect** it with what you will learn.

Active learning!!
To what should we p
when we subtract in
Subtraction in vertical form
They solved the problems of s
in vertical form in their math

Active learning!!

You can solve various problems by yourself, in a group, or with the whole class in this page.

Find the ?

What vegetable do you want to grow?

Problem Let's think about how to organize the data on vegetables you want to grow.

Let's think about how to organize data and how to represent them.

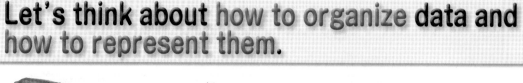

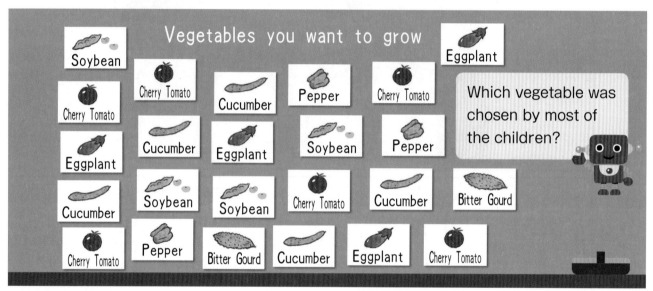

Vegetables you want to grow

Which vegetable was chosen by most of the children?

Want to know How to organize

1 There are 24 children in Hina's Class. Each child chose a vegetable that he or she wants to grow.

Let's look at the number of children who chose each vegetable.

① Let's write the number of children who chose each vegetable in the **table** below.

Way to see and think

It is easier to represent on the table by sorting vegetables.

Vegetables You Want to Grow

Vegetables	Cherry Tomatoes	Cucumbers	Soybeans	Eggplants	Peppers	Bitter Gourds
Number of Children	6					

② Represent the number of children who chose each vegetable by using ○ on the **graph** on the right.

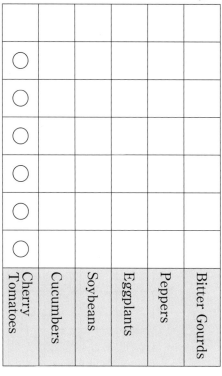

Vegetables You Want to Grow

The table is easier to read because the numbers of people who chose each vegetable are represented.

Nanami

Hiroto

The graph is better to read the difference in number.

③ Which vegetable was chosen by the most children? How many children chose it?

④ How many more children chose peppers than bitter gourds?

2 There are 15 children in Haruma's class.

The table below represents the numbers of milk teeth each child lost.

How can you make the difference in the numbers easier to read?

Number of Lost Milk Teeth

Name	Haruma	Shun	Tomomi	Takuma	Airi	Rin	Mana	Miyu
Number of Teeth	1	0	3	2	2	4	3	2

Name	Kana	Yuma	Toshiya	Mai	Momoka	Hiroki	Shota
Number of Teeth	1	3	2	4	3	2	1

① Represent the numbers by using ○ on the graph below.

Number of Lost Milk Teeth

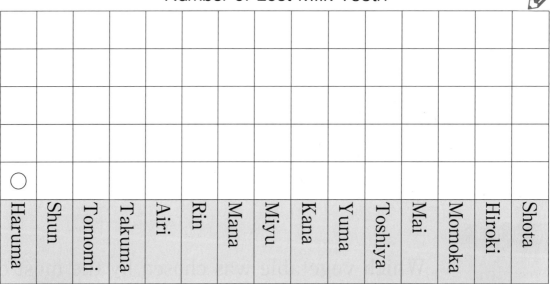

Nanami: There are a lot of children who lost 2 milk teeth.

Hiroto: If I make groups of the same number ...

Purpose Can we find an easier way to read the graph?

② Let's arrange the numbers of ○ in descending order.

Number of Lost Milk Teeth

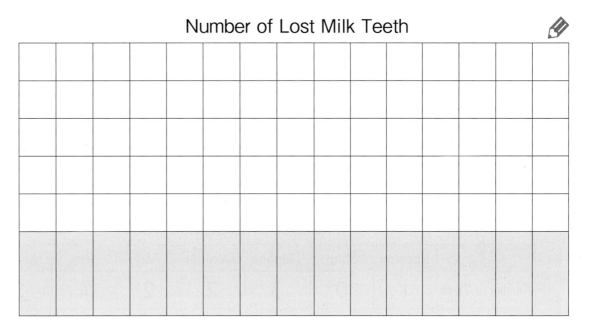

③ What is the most number of lost teeth and what is the least number?

④ Let's talk about how to read the difference in the numbers of lost teeth.

Daiki

Should I compare the numbers of ○?

I think it makes reading easier to write 1, 2, 3, … along the horizontal line of the graph.

Yui

⚲ Summary

It makes reading easier to arrange the numbers of ○ from the left of the graph in descending order and to write numbers along the horizontal line.

Let's **deepen.**

What is the most number of children who lost teeth?

Daiki

15

Deepen.

Utilize in life.

From another point of view

Want to know

The table below is the same as the one in **2** on page 14.

How many children lost the same number of milk teeth?

Number of Lost Milk Teeth

Name	Haruma	Shun	Tomomi	Takuma	Airi	Rin	Mana	Miyu
Number of Teeth	1	0	3	2	2	4	3	2

Name	Kana	Yuma	Toshiya	Mai	Momoka	Hiroki	Shota
Number of Teeth	1	3	2	4	3	2	1

Let's represent how many children lost the same number of milk teeth by using ○ on the graph.

Number of Lost Milk Teeth

0	1	2	3	4

Yui

Shun is the only child who lost no teeth.
So one ○ should be written in the column on 0 teeth.

Want to deepen

Compared to the graph in **2**, let's talk about the good points of each graph.

What you can do now

Can represent what I examined on a table or a graph.

1 Let's look at the weather in March shown below.

Weather in March

1	2	3	4	5	6	7	8	9	10	11
☀	☀	⛄	⛄	☁	⛄	☁	☂	☂	☂	☁

12	13	14	15	16	17	18	19	20	21	22
☀	☀	☀	☁	☁	☁	☀	☀	☂	☁	☂

23	24	25	26	27	28	29	30	31
☀	☀	☀	☁	☁	☂	☂	☁	☀

☀ Sunny ☁ Cloudy ☂ Rainy ⛄ Snowy

① Write the number of days in each weather condition in the table below.

Weather in March

Weather	Sunny	Cloudy	Rainy	Snowy
Number of days				

② Represent the number of days by using ○ on the graph on the right.

③ Which were more, sunny days or rainy days? By how many?

Weather in March

Sunny	Cloudy	Rainy	Snowy

Supplementary Problems p. 134

How long do we have to wait?

Problem   Let's think about how long they have to wait.

2 Time and Duration (I)

By **telling time** and **duration,** let's apply it into our life.

❶ Time and Duration

Want to explore How to represent time and duration

We went to explore the town.

The time we left school.

The time we arrived at the Fire Station.

1 Let's find out what they did in the town by telling time.

We have learned how to express time in the 1st grade.

How do the long hand and the short hand move?

A

The time we left school

B

The time we arrived at
the Fire Station

C

The time we left the
Fire Station

① Tell the **time** shown on each clock A, B, C, D, E, and F.

② Tell the **duration** it took the children to arrive at
the Fire Station after they left school.

The duration which
takes for the long hand to
move from | scale to the
next is called | **minute**.
The word minute can be
written as **min**.

If the long hand of the
clock moves |5 marks, then
|5 minutes have passed.

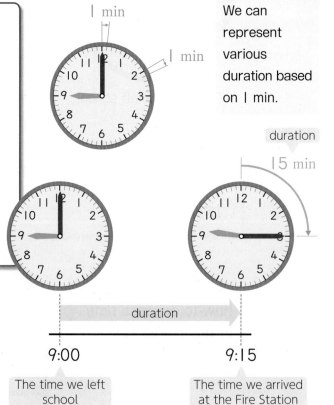

Way to see
and think

We can
represent
various
duration based
on | min.

| min

| min

duration

|5 min

duration

9:00

The time we left
school

9:|5

The time we arrived
at the Fire Station

D

The time we arrived at the flower shop

E

The time we left the flower shop

F

The time we came back to school

Want to represent

③ How much time did we spend at the flower shop?

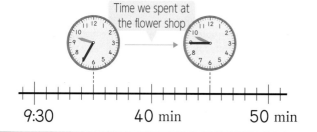

Time we spent at the flower shop

9:30   40 min   50 min

The duration which the long hand rotates completely around the clock is 60 minutes.

60 minutes is called 1 **hour**.

The word **hour** can be written as **hr**.

| 1 hour = 60 minutes |

1 hour

1 hour

It takes 1 hour for the short hand to move from 1 number to the next on the clock.

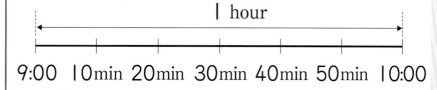

9:00 10min 20min 30min 40min 50min 10:00

Want to confirm

④ How long did it take from the time we left school to the time we came back to school?

morning

| | 0 | 1 | 2 | 3 | 4 | 5 | 6 | 7 | 8 | 9 | 10 | 11 | 12 (o'clock) |

(noon)

| 7 | 8 | 9 | 10 | 11 | 12 | | | | | | | 0 |

yesterday   today

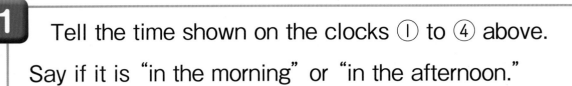

❷ The Duration in 1 Day

  Morning and afternoon

1   Tell the time shown on the clocks ① to ④ above.

Say if it is "in the morning" or "in the afternoon."

1   Talk about what you do in one day from the morning to the afternoon.

That's it! 💡 Morning · Afternoon

"a.m.," which stands for ante meridian, means before noon (morning).

9:00 in the morning can be written as 9:00 a.m.

"p.m.," which stands for post meridian, means after noon (afternoon).

2:00 in the afternoon can be written as 2:00 p.m.

Hiroto

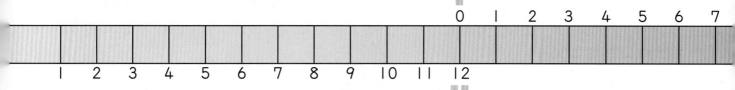

afternoon

```
          0   1   2   3   4   5   6   7
|   |   |   |   |   |   |   |   |   |   |   |
1   2   3   4   5   6   7   8   9  10  11  12
```

tomorrow

③

④

"12 o'clock midnight" is the same as "12 o'clock in the afternoon."
"12 o'clock in the afternoon" is the same as "12 o'clock midnight."

Want to explore

2 ▶ Let's see how many hours we have in one day.

Each day starts at 12 o'clock midnight.

The short hand rotates completely around the clock twice a day.

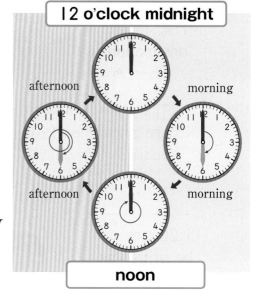

12 o'clock midnight

afternoon   morning

afternoon   morning

noon

The first complete rotation of the short hand makes ➡ morning ··· [ ] hours.

The next complete rotation of the short hand makes ➡ afternoon ··· [ ] hours.

1 day = 24 hours

That's it! 💡 Bus Schedule

The table below is the bus schedule.

Let's examine how a bus schedule is shown.

① The word neither "a.m." nor "p.m." is used.
Let's explain why.

In the table at the right, "時刻" means "hours," "行先" means "destination."

There are numbers that are larger than 12 in the columns below "時刻."

Hiroto

1:20 p.m. can be written as 13:20.

12:00 p.m. can be written as 0:00 a.m. next day.

行先 / 時刻	月～金 鴨江医療センター	鴨江大平台	三ヶ日	土・日祝 鴨江医療センター	鴨江大平台
6	36 51			41	
7	06 20 31 42 53	27		31 51	● 11 41
8	05 15 25 35 47 57	11 40		05 25 35 55	15 45
9	07 22 32 47 57	14 40		05 25 35 55	15 45
10	07 22 32 47 57	14 40		07 22 32 47 57	14 40
11	07 22 32 47 57	14 40		07 22 32 47 57	14 40
12	07 22 32 47 57	14 40		07 22 32 47 57	14 40
13	07 22 32 47 57	14 40		07 22 32 47 57	14 40
14	07 22 32 47 57	14 40		07 22 32 47 57	14 40
15	07 22 32 47 57	14 40		07 22 32 47 57	14 40
16	07 22 32 47 57	14 40		07 22 32 47 57	14 40
17	07 22 32 47 57	14 40	学23	07 22 32 47 57	14 40
18	07 22 32 47 57	14 40	学43	佐07 17 27 47 57	● 37
19	佐07 19 43 55	31		佐07 19 43 55	31
20	15 35 55	05 25		15 55	● 05 35
21	15	35		15	
22	00 46	25		46	

学：開校日のみ運行

富：富塚車庫止まり　　佐：佐鳴台団地止まり
●：佐鳴台団地経由　　東：東名経由 三ヶ日行

□囲みは超低床ノンステップ「オムニバス」運行予定 車両整備等により「オムニバス」で運行できない場合があります

a.m.	p.m.

0 1 2 3 4 5 6 7 8 9 10 11 12 (o' clock)

0 1 2 3 4 5 6 7 8 9 10 11 12 13 14 15 16 17 18 19 20 21 22 23 24 (o' clock)

What you can do now

☐ Understanding the relationship between day, hour, and minute.

1 Let's fill in the ☐ with numbers.

① 1 day = ☐ hours

② 60 minutes = ☐ hours

③ The duration for the long hand to rotate completely around the clock once is ☐ hour.

④ The short hand rotates completely around the clock ☐ times a day.

☐ Understanding the relationship between a.m. and p.m.

2 Let's fill in the ☐ with words.

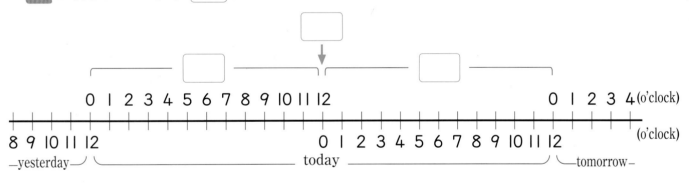

☐ Can tell time.

3 Let's write the time by using a.m. or p.m.

① The time I get up in the morning

② The time I go to school in the morning

③ The time I go to bed at night

Supplementary Problems ••••••••••▶ p. 134

Usefulness and efficiency of learning

1 Let's fill in the ☐ with numbers.

Understanding the relationship between day, hour, and minute.

① 1 hour = ☐ minutes

② The duration for the long hand to move 20 marks on the clock is called ☐ minutes.

③ The duration for the short hand to rotate completely around the clock once is ☐ hours.

2 Let's fill in the ☐ with words.

Understanding the relationship between a.m. and p.m.

① 12:00 p.m. can be written as 0:00 ☐.

② 0:00 p.m. can be written as ☐.

3 Let's draw the long hand of the clock that shows the time in the story.

Can tell time.

I went to school at 7:30 a.m.

I ate lunch at 0:20 p.m.

I studied until 2:25 p.m.

26

How many cookies were made altogether?

I made 12 cookies yesterday.

I made 23 cookies today.

?

How many cookies did I make altogether?

Problem Let's think about how many cookies she made altogether.

3 Addition and Subtraction of 2-digit Numbers
Let's think about how to calculate easily.

The number in each place is called a digit. 12 is a 2-digit number.

1 Addition

Want to think Addition of 2-digit numbers

Activity

1　I made 12 cookies yesterday and 23 cookies today. How many cookies are there altogether?

① Let's write a math expression to find the total number of cookies.

② Let's think about how many cookies there are altogether.

Can I find an easier way?

Daiki

Way to see and think

Let's think by using counters or blocks to represent cookies.

28

Nanami's idea

I replaced the cookies by counters and counted them by using sets of 10 counters.

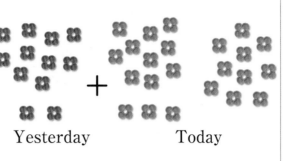

Yesterday   Today

Way to see and think

Sets of 10 are used.

Hiroto's idea

I replaced the cookies by ● and circled them in sets of 10 counters.

Yui's idea

I replaced the cookies by blocks for counting.

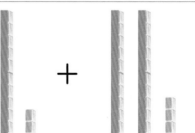

Want to compare

③ What is common among their ideas?

Want to think

④ Let's think about how to calculate $12 + 23$.

How should I calculate 2-digit numbers?

Daiki

Remember we made sets of 10 to calculate 1-digit numbers.

Yui

Way to see and think

You calculated the numbers in the ones place for $12 + 3$.

Y Purpose How can we add 2-digit numbers?

Daiki's idea

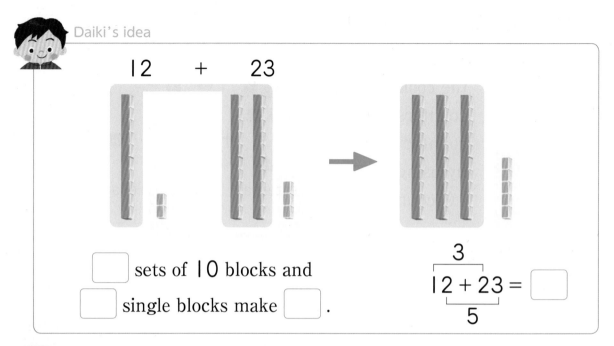

12 + 23

☐ sets of 10 blocks and
☐ single blocks make ☐.

$$3$$
$$12 + 23 = ☐$$
$$5$$

Yui's idea

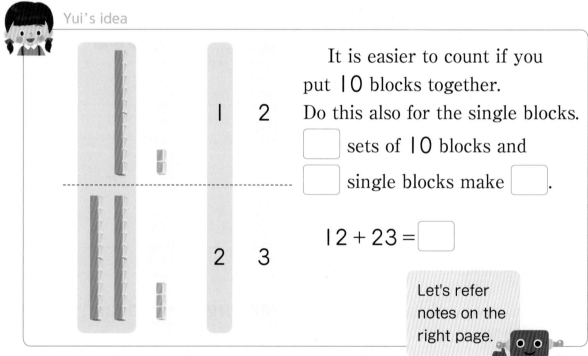

1   2

2   3

It is easier to count if you put 10 blocks together.
Do this also for the single blocks.

☐ sets of 10 blocks and
☐ single blocks make ☐.

$$12 + 23 = ☐$$

Let's refer notes on the right page.

⑤ Let's summarize your ideas in your notebook.

🧑 Summary

For adding 2-digit numbers, add the numbers in the ones place and then add the numbers in the tens place.

Notebook for thinking

Let's write the ideas and doubts that you had.

April 20

> I made 12 cookies yesterday and 23 cookies today.
> How many cookies are there altogether?

Math expression : 12 + 23

It seems that I can find the sum by using addition.

Purpose : How can I add 2-digit numbers?

〈My idea〉

I replaced the cookies by ● and circled them in sets of 10 counters for counting.

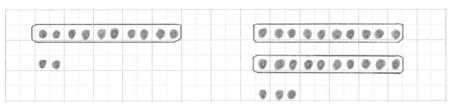

Answer : 35 cookies

〈Yui's idea〉

She replaced the cookies by blocks for counting.

It is easier to see sets of 10 by using blocks.

Write today's date.

Write the problem of the day that you must know.

Let's learn with the purpose.

Write your ideas or what you found about the problem.

Write the classmate's ideas you consider good.

❷ Subtraction

Subtraction of 2-digit numbers

Activity

1 Miku made 25 cookies. She gave 13 cookies to her brother.

How many cookies are left?

① Let's write a math expression to find the number of cookies left.

② Let's think about how many cookies are left.

Way to see and think

Let's think by replacing the cookies by counters or blocks.

Nanami's idea

I used counters to represent the cookies and then removed 13 counters.

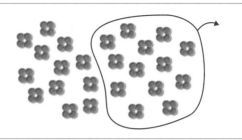

Hiroto's idea

I used ● to represent the cookies. I arranged them in sets of 10 counters and then removed 13 counters.

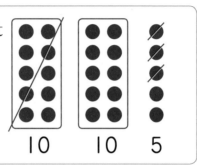

10   10   5

Way to see and think

Sets of 10 are used.

Yui's idea

I used blocks to represent the cookies and then removed 13 blocks.

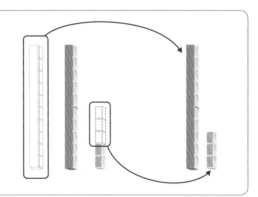

③ Let's think about how to calculate for finding the answer for 25 – 13.

Way to see and think

You calculated the numbers in the ones place for 25 – 3.

Can we subtract 2-digit numbers?

Daiki

We decomposed a 2-digit number into 10 and some in the 1st grade.

Yui

◯ Purpose   How can we subtract 2-digit numbers?

33

Daiki's idea

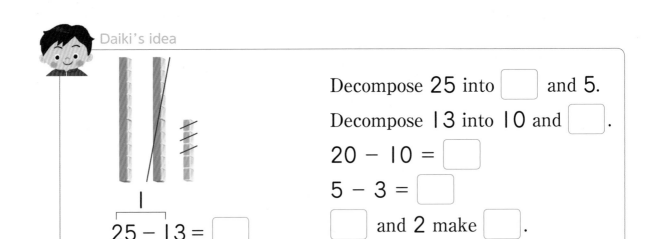

Decompose 25 into ☐ and 5.

Decompose 13 into 10 and ☐.

20 − 10 = ☐

5 − 3 = ☐

☐ and 2 make ☐.

25 − 13 = ☐

Yui's idea

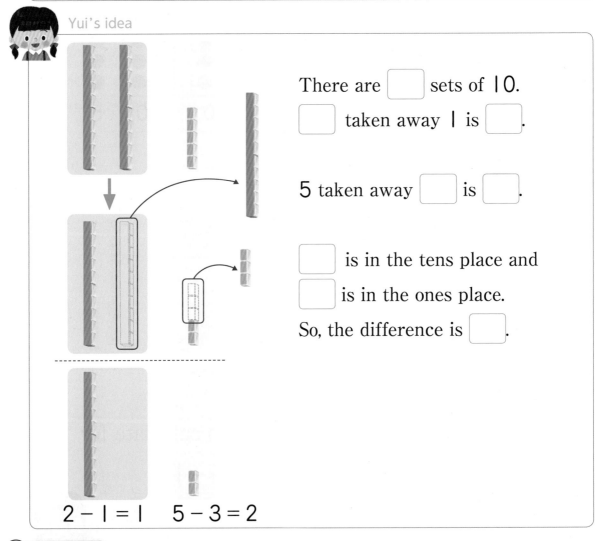

There are ☐ sets of 10.

☐ taken away 1 is ☐.

5 taken away ☐ is ☐.

☐ is in the tens place and

☐ is in the ones place.

So, the difference is ☐.

2 − 1 = 1   5 − 3 = 2

🌸 Summary

For subtracting 2-digit numbers, subtract the numbers in the ones place and then subtract the numbers in the tens place.

4 Addition in Vertical Form

Let's think about the meaning of addition and how to add.

1 Addition of 2-digit numbers

Want to know Addition in vertical form

1 Tulips have bloomed. There are 24 red tulips and 13 yellow tulips.

How many tulips are there altogether?

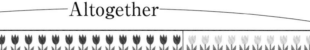

Altogether

24 red tulips   13 yellow tulips

① Let's write a math expression.

24 + 13 can be written vertically by aligning the digits of the numbers according to their places. Such algorithm is called addition in **vertical form**.

Want to think

② Let's think about how to add 24 + 13 in vertical form.

Purpose How do we add in vertical form?

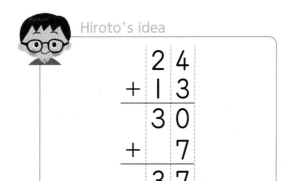

Hiroto's idea

```
   2 4
 + 1 3
   3 0
 +   7
   3 7
```

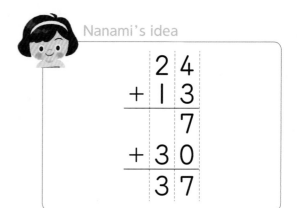

Nanami's idea

```
   2 4
 + 1 3
     7
 + 3 0
   3 7
```

Addition algorithm for 24 + 13 in vertical form

Tens	Ones

```
   2 4
 + 1 3
```
→
```
   2 4
 + 1 3
   3 7
```

2 + 1 = 3   4 + 3 = 7

Align the digits of the numbers according to their places.

Add the numbers in the ones place and then add the numbers in the tens place.

Math Sentence : 24 + 13 = 37   Answer : 37 tulips

Want to confirm

1 Let's think about how to add 17 + 30 in vertical form.

```
   1 7
 + 3 0
```

Want to try

2 Let's calculate the following in vertical form.

① 31 + 57   ② 26 + 43   ③ 15 + 62   ④ 65 + 31

⑤ 18 + 40   ⑥ 32 + 20   ⑦ 50 + 36   ⑧ 20 + 70

2 Let's think about how to add $2 + 41$ in vertical form.

① Which is the correct way of writing the numbers?

Daiki's idea

$$\begin{array}{r} 2 \\ + \, 4 \;\; 1 \\ \hline \end{array}$$

Yui's idea

$$\begin{array}{r} 2 \\ + \, 4 \;\; 1 \\ \hline \end{array}$$

Tens	Ones

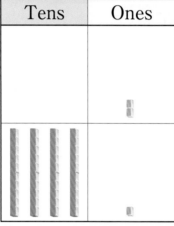

Let's think about the approximate answer to $2 + 41$.

② Let's add in vertical form.

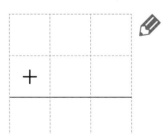

Summary

For adding in vertical form, align the digits of the numbers according to their places and then add the numbers in the same places.

3 Let's calculate the following in vertical form.

① $4 + 23$ ② $7 + 82$ ③ $91 + 8$ ④ $65 + 3$

3 There are 38 picture books and 27 reference books in Sakura's classroom.

How many books are there altogether?

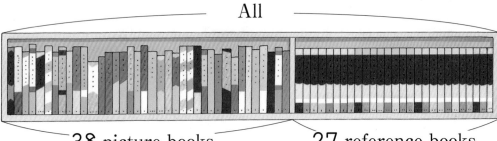

All

38 picture books   27 reference books

① Let's write a math expression.

Tens	Ones

② Let's think about how to add in vertical form.

$$\begin{array}{r} 3\ 8 \\ +\ 2\ 7 \\ \hline \end{array}$$

Nanami

What is the difference from the calculation of 24 + 13?

In the ones place, 8 + 7 = 15. So ...

Hiroto

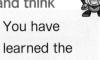

Way to see and think

You have learned the calculation of 8 + 7 in the 1st grade.

Purpose When the answer to the addition of the ones place is more than 10, how can we add in vertical form?

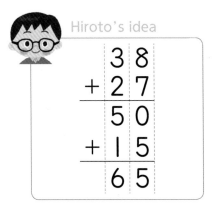

Hiroto's idea

```
  3 8
+ 2 7
  5 0
+ 1 5
  6 5
```

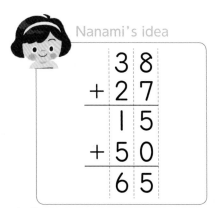

Nanami's idea

```
  3 8
+ 2 7
  1 5
+ 5 0
  6 5
```

Can you not reduce the number of rows?

If the sum is more than 10 when we add the numbers in a place, the set of 10 is moved to the next higher place. This is called **carrying** or **regrouping**.

Addition algorithm for 38 + 27 in vertical form		

Tens	Ones

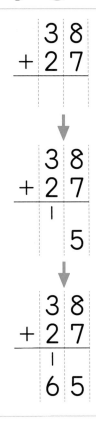

Carry

```
  3 8
+ 2 7
```
Align the digits of the numbers according to their places.
Add the numbers in the ones place first.

```
  3 8
+ 2 7
  1
    5
```

Ones Place

$8 + 7 = 15$

The ones place is ☐.
Carry 1 ten to the tens place.

```
  3 8
+ 2 7
  1
  6 5
```

Tens Place

1 ten was carried, so

$3 + 2 + 1 = 6$

The tens place is ☐.

Math Sentence : $38 + 27 = 65$ Answer : 65 books

Summary

If a set of 10 is made when we add the numbers in the ones place, carry 10 ones to the tens place as 1 ten.

4 Let's calculate the following in vertical form.

① 27 + 65　　　② 48 + 34

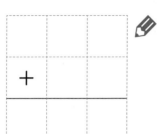　　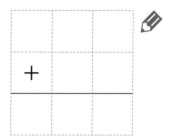

 5 Let's calculate the following in vertical form.

① 28 + 16　② 47 + 27　③ 59 + 37　④ 15 + 56

⑤ 43 + 38　⑥ 18 + 78　⑦ 24 + 19　⑧ 49 + 13

That's it! How to write the carried number in vertical form

There are many ways of writing.

```
    1
  3 8        3 8
+ 2 7      + 2①7
  6 5        6 5
```

Think of the ways of doing this so you don't forget the carried 1.

Daiki

40

4 Find the mistakes in the following processes.

Let's talk about how to calculate correctly.

① 27 + 53

```
  2 7
+ 5 3
─────
  7 0
```

```
+
```

Be careful when carrying.

Hiroto

② 35 + 6

```
  3 5
+   6
─────
  9 5
```

```
+
```

③ 7 + 23

```
    7
+ 2 3
─────
  2 0
```

```
+
```

6 The answer of the addition sentence below is 50.

Let's make the math sentence.

□ + □ = 50

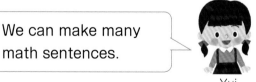
We can make many math sentences.

Yui

7 Let's calculate the following in vertical form.

① 72 + 18 ② 35 + 45 ③ 16 + 24 ④ 33 + 17

⑤ 54 + 7 ⑥ 77 + 9 ⑦ 6 + 89 ⑧ 5 + 15

Want to know

1 There are 38 strawberries in a box and 16 strawberries in a basket.

How many strawberries are there altogether?

① Let's put the strawberries in the basket into the box.

augend		addend		sum
38	+	16	=	

② Let's put the strawberries in the box into the basket.

augend		addend		sum
16	+	38	=	

In addition, the answers are the same even when the order of the augend and the addend are changed.

$$38 + 16 = 16 + 38$$

Want to confirm

1 Let's find the math expressions with the same answer.

24 + 31   70 + 9   37 + 21   9 + 62

Can you find some without calculation?

21 + 37   62 + 9   72 + 11   31 + 24

2   I had 32 marbles. I received 7 marbles from my brother and 3 marbles from my sister.

   How many marbles do I have altogether?

① Let's write a math expression.

② Let's think about how to calculate.

Daiki's idea

   Since I received 7 marbles from my brother,
   $32 + 7 = 39$.
   Since I received 3 marbles from my sister,
   $39 + 3 = 42$.

Nanami's idea

   I combined the marbles I received from my brother and sister,
   $7 + 3 = 10$.
   Because I already had 32 marbles,
   $32 + 10 = 42$.

   In addition, the answers are the same even when the order of operation is changed. $32 + (7 + 3)$

   $(32 + 7) + 3 = 32 + (7 + 3)$

   ( ) is the symbol that indicates we calculate first.   ②

① ②

2   Let's find an easier way to add.

① $45 + 18 + 2$ ② $58 + 13 + 27$

③ $68 + 13 + 12$ ④ $44 + 37 + 6$

Which combination of numbers makes addition simpler?

What you can do now

☐ Understanding the meaning of the way to add in vertical form.

1 Let's summarize how to add $67 + 28$ in vertical form.

(1) In the ones place, $7 + 8$ makes 15.

The ones place is ☐.

Carry ☐ ten to the tens place.

(2) In the tens place, $6 + 2 + ☐ = 9$

(3) The answer is ☐.

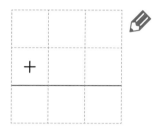

☐ Can add in vertical form.

2 Let's calculate the following in vertical form.

① $36 + 32$ ② $43 + 34$ ③ $2 + 55$

④ $40 + 47$ ⑤ $38 + 25$ ⑥ $57 + 19$

⑦ $35 + 58$ ⑧ $17 + 43$ ⑨ $18 + 9$

⑩ $49 + 4$ ⑪ $8 + 47$ ⑫ $5 + 75$

☐ Can find the answers by making addition expressions.

3 If you buy a chewing gum for 42 yen and a candy for 55 yen, what will be the total cost?

☐ Understanding the rules of addition.

4 Let's calculate the following in vertical form. Then, change the order of the numbers and find the sum. Confirm your answers.

① $73 + 15$ ② $24 + 70$ ③ $7 + 29$

☐ Can find an easier way to add.

5 Let's find an easier way to add.

① $56 + 22 + 8$ ② $4 + 37 + 26$

Supplementary Problems p. 135

Usefulness and efficiency of learning

1 Find the mistakes in the following processes.
Let's write the correct answer in the ( ).

Understanding the meaning of the way to add in vertical form.

①
```
    2 7
  + 4 3
  -------
    6 0
```
( )

②
```
      6
  + 3 5
  -------
    9 5
```
( )

2 Let's calculate the following in vertical form.

Can add in vertical form.

① 14 + 63   ② 45 + 24   ③ 30 + 56
④ 42 + 29   ⑤ 36 + 47   ⑥ 19 + 65
⑦ 22 + 18   ⑧ 54 + 16   ⑨ 33 + 57
⑩ 67 + 3   ⑪ 69 + 8   ⑫ 3 + 29

3 In Itsuki's school, there are 2 classes for the 2nd grade. There are 31  children in the 1st class and 29 children in the 2nd class. How many children are there in the 2nd grade altogether?

Can find the answers by making addition expressions.

4 Let's calculate the following in vertical form. Then, change the order of the numbers and find the sum. Confirm your answers.

Understanding the rules of addition.

① 39 + 47   ② 54 + 36

5 Let's find an easier way to add.

Can find an easier way to add.

① 17 + 38 + 23   ② 32 + 23 + 27

Find the ? How many are left?

Problem Let's think about how many pieces are left.

Let's think about the meaning of subtraction and how to subtract.

1 Subtraction of 2-digit numbers

Want to know   Subtraction in vertical form

1 Minato and his friends had 38 pieces of heart origami.

They gave 12 pieces to the 1st grade children.

How many pieces are left?

38 pieces

12 pieces given   the number of pieces left

① Let's write a math expression.

Want to think

② Let's think about how to subtract 38 − 12 in vertical form.

$$\begin{array}{r} 3\ 8 \\ -\ 1\ 2 \\ \hline \end{array}$$

Way to see and think

Remember the addition in vertical form.

Ⓨ Purpose   How do we subtract in vertical form?

Subtraction algorithm for 38 − 12 in vertical form

Tens	Ones

```
  3 8
－ 1 2
```
→
```
  3 8
－ 1 2
  2 6
```
3−1=2 8−2=6

Align the digits of the numbers according to their places.

Subtract the numbers in the ones place and then subtract the numbers in the tens place.

Math Sentence : 38 − 12 = 26   Answer : 26 pieces

Want to confirm

1 Let's think about how to subtract 47 − 25 in vertical form.

```
    4 7
－  2 5
```

Want to try

2 Let's calculate the following in vertical form.

① 76 − 32   ② 59 − 45   ③ 36 − 24

④ 49 − 13   ⑤ 97 − 76   ⑥ 66 − 25

2 Let's think about how to calculate the following in vertical form.

① 34 − 14   ② 68 − 64   ③ 54 − 40

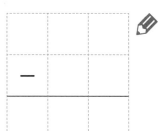   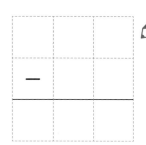   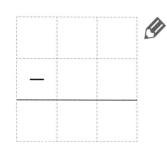

3 Let's calculate the following in vertical form.

① 29 − 6   ② 48 − 8

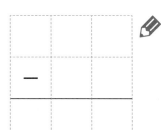   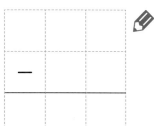

Let's think about the approximate answer to 29 − 6.

Summary

For subtracting in vertical form, align the digits of the numbers according to their places and then subtract the numbers in the same places.

4 Let's calculate the following in vertical form.

① 98 − 18   ② 43 − 42   ③ 30 − 20

④ 58 − 5   ⑤ 74 − 2   ⑥ 85 − 5

3 There were 45 sheets of origami paper. I used 27 sheets.

How many sheets are left?

① Let's write a math expression.

② Let's think about how to calculate in vertical form.

```
    4  5
 −  2  7
```

Let's think about the approximate answer.

Hiroto

Tens	Ones

What is the difference from the calculation of 38−12?

Nanami

In the ones place, we cannot do 5−7 ...

Daiki

Purpose When the subtraction of the ones place cannot be done, how can we subtract in vertical form?

Way to see and think

You have learned the calculation of 15−7 in the 1st grade.

As shown below, 1 ten is moved from the tens place to the ones place as 10 ones. This is called **borrowing** or **regrouping**.

Subtraction algorithm for $45 - 27$ in vertical form

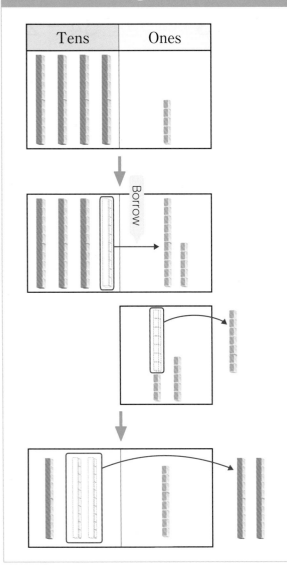

$$
\begin{array}{r}
4\ 5 \\
-\ 2\ 7 \\
\hline
\end{array}
$$

Align the digits of the numbers according to their places.

$$
\begin{array}{r}
{\scriptstyle 3}\ \ {\scriptstyle 10} \\
\not{4}\ 5 \\
-\ 2\ 7 \\
\hline
8
\end{array}
$$

Ones Place

Borrow 1 ten from the tens place as 10 ones, so $15 - 7 = 8$.

The ones place of the answer becomes ☐.

$$
\begin{array}{r}
{\scriptstyle 3}\ \ {\scriptstyle 10} \\
\not{4}\ 5 \\
-\ 2\ 7 \\
\hline
1\ 8
\end{array}
$$

Tens Place

1 ten has been moved to the ones place, so $3 - 2 = ☐$.

The tens place of the answer becomes ☐.

Math Sentence : $45 - 27 = 18$ Answer : 18 sheets

🗨 Summary

When the subtraction of the ones place cannot be done, borrow 1 ten from the tens place as 10 ones.

Be careful when borrowing.

4 Let's calculate the following in vertical form.

① 53 − 36　② 70 − 43　③ 34 − 26

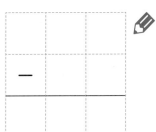　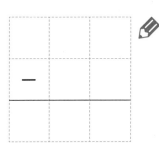　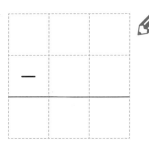

Want to confirm

5 Let's calculate the following in vertical form.

① 41 − 19　② 72 − 33　③ 81 − 16　④ 66 − 28

⑤ 70 − 56　⑥ 40 − 24　⑦ 50 − 33　⑧ 80 − 48

⑨ 26 − 18　⑩ 54 − 45　⑪ 73 − 67　⑫ 90 − 89

Want to think　(2-digit number) − (1-digit number) in vertical form

6 Let's think about how to subtract 35 − 8 in vertical form.

Way to see and think

Align the digits of the numbers.

Want to confirm

7 Let's calculate the following in vertical form.

① 92 − 8　② 51 − 9　③ 40 − 7　④ 60 − 3

52

Want to solve

1 There were 34 children in the classroom.

15 of them went outside to play.

How many are left in the classroom?

34 children in the classroom at first

15 children went outside.   ☐ children were left.

① Let's find the answer.

minuend	subtrahend	difference
34	− 15	=

② If the 15 children who went outside come

back to the classroom, how many will be there?

difference	subtrahend	minuend
19	+ 15	=

This method can be used to confirm the answer in subtraction.

Want to confirm

 Let's calculate the following. And then confirm your

answers.

① 76 − 51 ② 32 − 26 ③ 45 − 8 ④ 50 − 7

What you can do now

☐ Understanding the meaning of the way to subtract in vertical form.

1 Let's summarize how to subtract $73 - 26$ in vertical form.

(1) Borrow ☐ ten from the tens place, so that the ones place will become ☐ $- 6 =$ ☐ .

(2) In the tens place, ☐ $- 2 =$ ☐ .

(3) The answer is ☐ .

☐ Can subtract in vertical form and confirm the answer.

2 Let's calculate the following in vertical form. And confirm the answer.

① $58 - 32$ ② $66 - 23$ ③ $33 - 11$

④ $28 - 12$ ⑤ $87 - 19$ ⑥ $63 - 24$

⑦ $42 - 13$ ⑧ $54 - 26$ ⑨ $80 - 17$

⑩ $50 - 49$ ⑪ $34 - 27$ ⑫ $44 - 38$

☐ Can find the answers by making subtraction expressions.

3 Let's answer the following.

① Koharu had 32 candies. She gave 14 of them to her brother. How many candies are left?

② In the ball-toss game, the Red Team got 83 balls into the basket and the White Team got 79 balls. Which team got more and by how many?

Supplementary Problems ● ● ● ● ● ● ● ● → p. 136

Usefulness and efficiency of learning

1 Find the mistakes in the following processes.

Let's write the correct answer in the ( ).

Understanding the meaning of the way to subtract in vertical form.

①
```
  7 2
- 4 7
-----
  3 5
```
( )

②
```
  6 5
- 4 3
-----
  1 2
```
( )

③
```
  5 8
-   3
-----
  2 8
```
( )

2 Let's calculate the following in vertical form. And confirm the answer.

Can subtract in vertical form and confirm the answer.

① 67 − 42
② 59 − 30
③ 88 − 66
④ 96 − 16
⑤ 90 − 38
⑥ 80 − 47
⑦ 70 − 35
⑧ 60 − 45
⑨ 82 − 58
⑩ 56 − 27
⑪ 63 − 46
⑫ 31 − 23
⑬ 47 − 19
⑭ 82 − 7
⑮ 64 − 8

3 Let's answer the following.

Can find the answers by making subtraction expressions.

① The pencil is 53 yen and the eraser is 90 yen. Which is more expensive and by how much?

② There are 71 children in the 2nd grade at Yuna's school. 39 of them are girls. How many boys are there?

4 Let's fill in the ☐ with numbers.

```
  ☐ 4
- 1 ☐
-----
  1 9
```

Let's **deepen.**

I want to make problems like **4** on the left.

Daiki

Deepen.

Number in leaves eaten by worms

(Example)

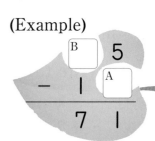

Want to try

What numbers were eaten by the worms? Let's fill in each ☐ with the number.

- A is the number that makes $5 - \boxed{A} = 1$, so A = 4.

- B is the number that makes $\boxed{B} - 1 = 7$, so B = 8.

```
   9  1
-  D  7
─────────
   6  C
```

```
   6  E
-  1  2
─────────
 F  8
```

Want to deepen

Make more problems like the ones above. Exchange them with your friends and solve them.

(How to Make Problems)

(1) Solve a problem correctly.

(2) Decide what numbers to replace with ☐.

(3) Do the problem yourself to confirm if it can be solved.

(Example 1)

```
   3  8
+  2  6
──────
   6  4
```

```
   ☐  8
+  2  ☐
──────
   6  4
```

(Example 2)

```
   8  7
-  2  9
──────
   5  8
```

```
   8  ☐
-  ☐  9
──────
   5  8
```

How many chicks are there?

Problem Let's think about how to represent large numbers.

6 Let's explore how to represent numbers and their structure.

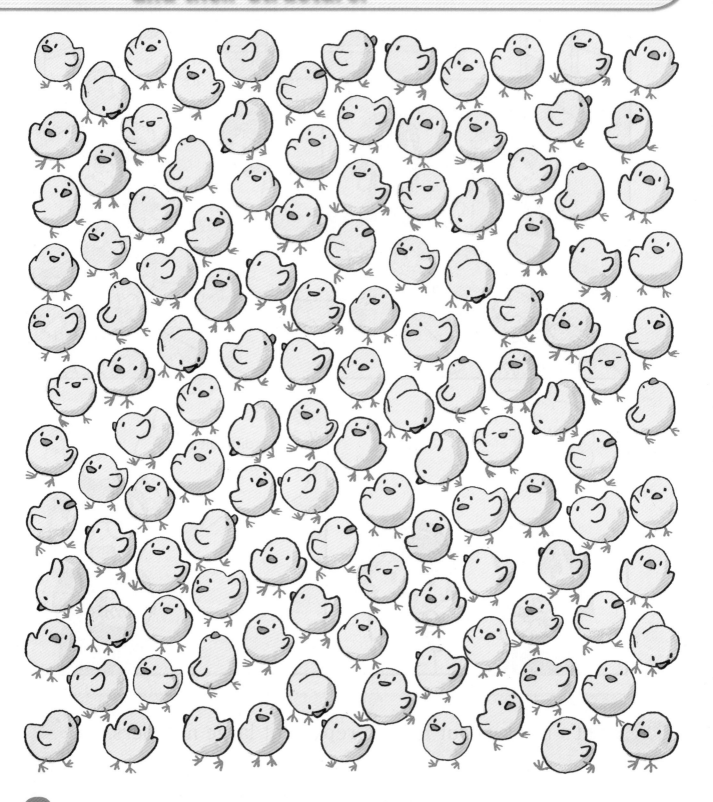

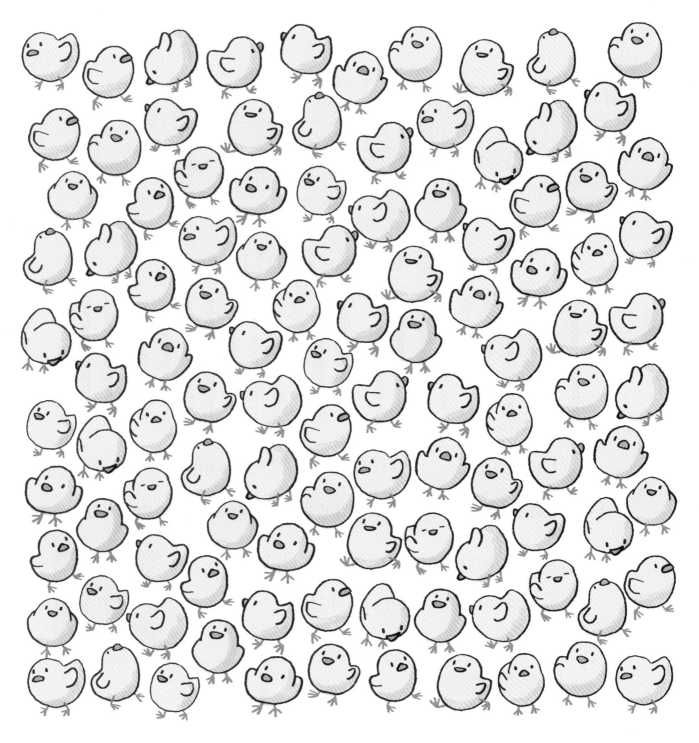

1 Numbers larger than 100

Want to explore

Can you find an easier way to count?

1 How many chicks are there altogether?

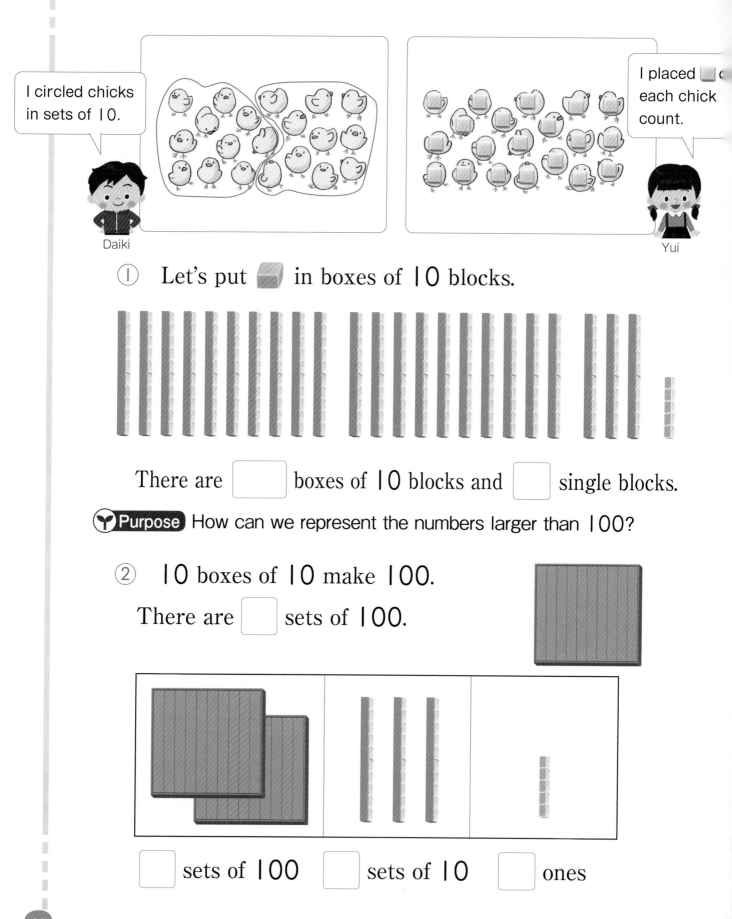

I circled chicks in sets of 10.

Daiki

I placed ▨ each chick count.

Yui

① Let's put ◻ in boxes of 10 blocks.

There are ☐ boxes of 10 blocks and ☐ single blocks.

Ⓨ Purpose How can we represent the numbers larger than 100?

② 10 boxes of 10 make 100.

There are ☐ sets of 100.

☐ sets of 100   ☐ sets of 10   ☐ ones

🌸Summary

Two sets of 100 is **two hundred**. Two hundred, thirty, and five is represented as **two hundred thirty-five** and is written as 235.

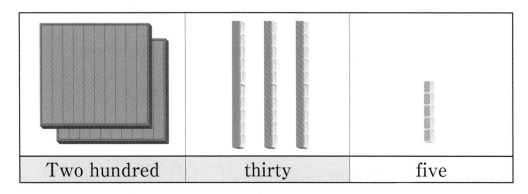

Two hundred	thirty	five

Hundreds Place	Tens Place	Ones Place
2	3	5

The position of 2 in 235 is called the **hundreds place**.

Want to confirm

 1 How many are there altogether?

①

Hundreds	Tens	Ones

Hundreds Place	Tens Place	Ones Place

②

Hundreds	Tens	Ones

Hundreds Place	Tens Place	Ones Place

2 Let's read the following numbers.

① 379   ② 516   ③ 847   ④ 136

 3 Let's write the following numbers.

① The number when seven hundred, thirty, and four are added.

② The number when one hundred, fifty, and seven are added.

③ The sum of 4 sets of 100, 9 sets of 10, and 5 ones.

④ The sum of 6 sets of 100, 1 set of 10, and 1 one.

Want to think

2 How many pencils are there?

Hundreds	Tens	Ones

Hundreds Place	Tens Place	Ones Place

 sets of 100 ☐ sets of 10 ☐ ones

The number of pencils is written as **307** and read as three hundred seven.

4 How many ▨ are there altogether?

①

Hundreds	Tens	Ones

Hundreds Place	Tens Place	Ones Place

The sum of 200 and 30.

②

Hundreds	Tens	Ones

Hundreds Place	Tens Place	Ones Place

The sum of 3 sets of 100.

 5 Let's read the following numbers.

① 820　② 160　③ 408　④ 900

6 Let's write the following numbers.

① seven hundred forty　② eight hundred ten　③ one hundred twenty

④ five hundred eight　⑤ one hundred one　⑥ six hundred

 7 Let's fill in each ☐ with a number.

① The sum of 5 sets of 100 and 6 ones is ☐.

② 860 is the sum of ☐ sets of 100 and ☐ sets of 10.

3 How many • are there in the diagram below?

① Circle the sets of 100 dots.

② How many sets of 100 are there?

10 sets of 100 is **one thousand** and is written as 1000.

 8 Let's write the following numbers by using the line of number below.

① What is the number that is added to 900 to get 1000?

② What is the number that is 10 smaller than 1000?

③ What is the number that is 1 smaller than 1000?

Way to see and think

Let's think about what the smallest scale on the line represents.

0   100   200   300   400   500

Want to try

9 Let's fill in each ☐ with a number.

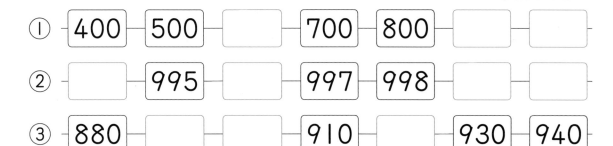

① 400 — 500 — ☐ — 700 — 800 — ☐ — ☐

② ☐ — 995 — ☐ — 997 — 998 — ☐ — ☐

③ 880 — ☐ — ☐ — 910 — ☐ — 930 — 940

10 Let's write the numbers in the ☐ that each ↑ is pointing at.

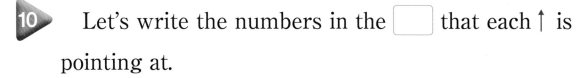

560　　570　　580　　590　　600　　610　　620

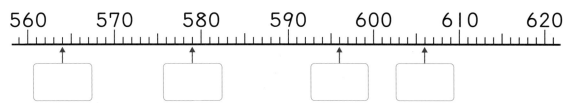

11 Let's draw ↑ at the scale on the line of number below that represents each number.

① 450　　② 680　　③ 990

12 Let's write the following numbers.

① The number that is 300 larger than 500.

② The number that is 200 smaller than 700.

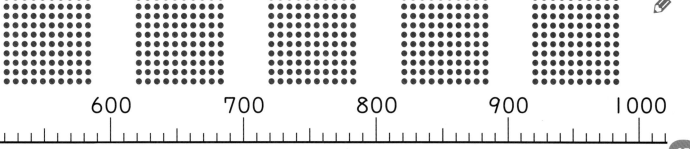

600　　　700　　　800　　　900　　　1000

4 Let's examine the number 230.

① There are **2** sets of 100.

How many sets of 10 are there?

② How many sets of 10 are there in all?

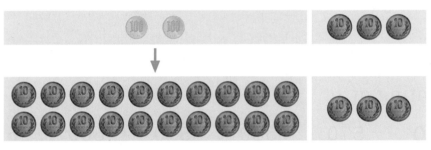

$$230 \begin{cases} 200 \rightarrow 20 \text{ sets of } 10 \\ 30 \rightarrow 3 \text{ sets of } 10 \end{cases} \boxed{} \text{ sets of } 10$$

Way to see and think

Let's change all to 10-yen coins to know the amount.

Want to try

 13 What is the number that is the sum of 14 sets of 10?

Way to see and think

The number that is the sum of 10 sets of 10 is 100.

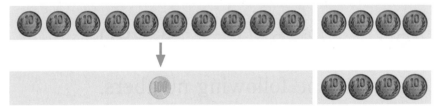

$$14 \text{ sets of } 10 \begin{cases} 10 \text{ sets of } 10 \rightarrow 100 \\ 4 \text{ sets of } 10 \rightarrow 40 \end{cases} \boxed{}$$

Want to confirm

 14 Let's fill in each $\boxed{}$ with a number.

① **560** is the sum of $\boxed{}$ sets of 10.

② $\boxed{}$ is the sum of **37** sets of 10.

Want to compare

The table on the right shows the numbers of plastic bottle caps that were collected by the 1st and the 2nd classes.

Let's compare the numbers of caps. Which class collected more caps?

1st Class	290 caps
2nd Class	288 caps

① Let's compare the numbers by writing numbers in the table on the right.

	Hundreds Place	Tens Place	Ones Place
1st Class			
2nd Class			

Way to see and think

In which place can you compare the numbers?

> and < are signs to represent larger and smaller for comparing sizes.

When the size is the same, = is used.

4 > 2

4 is larger than 2.

3 = 3

3 has the same size as 3.

2 < 4

2 is smaller than 4.

Want to confirm

Let's write > or < in each ☐.

① 238 ☐ 253

220   230   240   250   260

② 769 ☐ 764

750   760   770   780

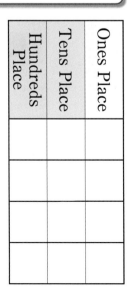

Hundreds Place	Tens Place	Ones Place

1 You want to buy a chewing gum for 50 yen and a candy for 80 yen. What is the total cost?

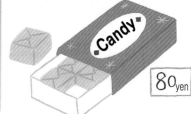

Math Sentence :

Way to see and think

Let's think about how many sets of ten ...

Want to try

1 You have 130 yen. If you buy a chocolate for 70 yen, how much will be left?

Math Sentence :

Way to see and think

Let's think about how many sets of 10 there are in 130.

Want to confirm

2 Let's calculate the following.

① 70 + 50   ② 90 + 20   ③ 60 + 60

④ 130 − 50   ⑤ 140 − 80   ⑥ 160 − 90

What you can do now

☐ Understanding how to represent numbers.

1 How many sheets of colored paper are there?

☐ Understanding the structure of numbers.

2 Look at 480 and let's fill in each ☐ with a number.

① 4 in the hundreds place means that 4 has a value of ☐ .

② 480 is the sum of ☐ sets of 10.

☐ Understanding the sizes of 3-digit numbers.

3 Let's write > or < in each ☐ .

① 718 ☐ 781   ② 555 ☐ 559

③ 310 ☐ 301   ④ 615 ☐ 610

☐ Can use sets of 10 to calculate.

4 Let's calculate the following.

① 40 + 90   ② 70 + 70

③ 130 − 60   ④ 150 − 80

Supplementary Problems ▶ p. 138

Usefulness and efficiency of learning

1 Let's fill in each ☐ with a number.

Understanding how to represent numbers.

① The number when 500, 10, and 9 are added is ☐.

② The number that is the sum of 2 sets of 100 and 5 ones is ☐.

③ 964 is the sum of ☐ sets of 100, ☐ sets of 10, and ☐ ones.

2 740 is represented as shown below.

Let's fill in each ☐ with a number.

Understanding the structure of numbers.

740 is the number that is 60 smaller than ☐.	740 is the number when ☐ and 40 are added.	740 is the number that is the sum of ☐ sets of 10.

3 The number in the ones place of the White Team's score is still unknown. Takuto says that the Red Team has won. Why does he say so?

Understanding the sizes of 3-digit numbers.

Red   8 5 1   White   8 4

4 Let's answer the following.

Can use sets of 10 to calculate.

① You have 30 pieces of chocolate. Then you get 40 more pieces. How many pieces do you have altogether?

② There are 170 sheets of paper. You use 90 sheets. How many sheets are left?

How many plastic bottles?

Panel 1: I participated in cleaning the town last month. I picked up 74 plastic bottles.

Panel 2: 65 plastic bottles this month.

Panel 3: How many bottles did you pick up altogether?

Panel 4: I picked up more than 100 bottles. ....

Problem Let's think about how to calculate large numbers.

7 Addition and Subtraction of Large Numbers
Let's think about the meaning of calculation and how to calculate.

1 Addition with 3-digit answers

Want to solve Addition with carrying to the hundreds place

1 We participated in cleaning the town. We picked up 74 plastic bottles last month and 65 plastic bottles this month. How many bottles did we pick up altogether?

① Let's write a math expression.

About how many did you pick up?

Want to think

② Let's think about how to calculate.

Nanami's idea

74 ··· 70 + 4
65 ··· 60 + 5

130 and 9
make 139

Hiroto's idea

```
    7 4
+   6 5
------
      9
  1 3 0
------
  1 3 9
```

Way to see and think

Remember the addition of 2-digit numbers in vertical form.

Want to communicate

③ Let's explain how to add 74 + 65 in vertical form.

```
    7 4
+   6 5
------
```

Addition algorithm for 74 + 65 in vertical form

Tens	Ones
∥∥∥∥∥∥∥	│
∥∥∥∥∥∥	│

Hundreds	Tens	Ones
▨	∥∥∥	│

Carry

$$
\begin{array}{r}
7\ 4 \\
+\ 6\ 5 \\
\hline
\end{array}
$$

→

$$
\begin{array}{r}
7\ 4 \\
+\ 6\ 5 \\
\hline
9 \\
\end{array}
$$

→

$$
\begin{array}{r}
7\ 4 \\
+\ 6\ 5 \\
\hline
1\ 3\ 9 \\
\end{array}
$$

Align the digits of the numbers according to their places.

Ones Place

$4 + 5 = 9$

The number in the ones place is 9.

Tens Place

$7 + 6 = 13$

The tens place is 3. Carry 10 tens to the hundreds place as 1 hundred.

Math Sentence : $74 + 65 = 139$ Answer : 139 bottles

Want to confirm

1   Let's calculate the following in vertical form.

① $93 + 86$ ② $63 + 71$ ③ $67 + 80$ ④ $20 + 90$

2 Let's explain how to add $48 + 87$ in vertical form.

	4	8
+	8	7

Addition algorithm for $48 + 87$ in vertical form

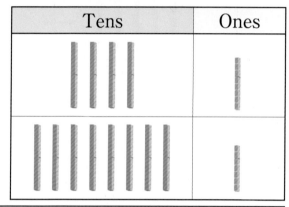

Way to see and think

You can add by carrying as you have learned.

Align the digits of the numbers according to their places.

Ones Place

$8 + 7 = 15$

The number in the ones place is 5.
Carry 10 ones to the tens place as 1 ten.

Tens Place

$4 + 8 + 1 = 13$

The tens place is 3.
Carry 10 tens to the hundreds place as 1 hundred.

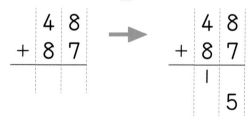

2 Let's calculate the following in vertical form.

① 35 + 96   ② 88 + 44   ③ 36 + 89

④ 54 + 67   ⑤ 51 + 69   ⑥ 32 + 78

3 Let's think about how to calculate the following in vertical form.

① 37 + 67   ② 6 + 97

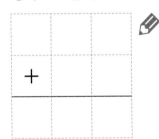   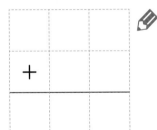

Will 10 tens be carried?

③ 15 + 85   ④ 9 + 91

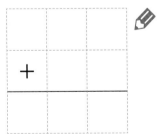   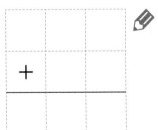

 3 Let's calculate the following in vertical form.

① 27 + 78   ② 32 + 69

③ 51 + 49   ④ 58 + 42

⑤ 8 + 96   ⑥ 93 + 7

1

There were 400 sheets of paper.

A number of sheets were placed on them.

How many sheets of paper are there altogether?

① 300 sheets of paper were placed. How

many sheets are there altogether?

Math Expression :

Answer : [ ] sheets

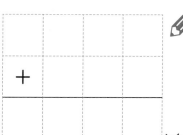

② 300 more sheets of paper are

placed on top of the stack. How many

sheets will be there altogether?

Way to see and think

Let's think by using bundles of 100.

Math Expression :

Answer : [ ] sheets

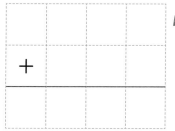

Want to confirm

1 Let's calculate the following.

① 100 + 400   ② 200 + 600   ③ 700 + 200

④ 900 + 100   ⑤ 600 + 400   ⑥ 200 + 800

2 Let's calculate the following in vertical form.

① 628 + 7   ② 234 + 57

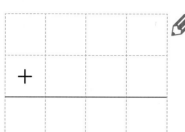
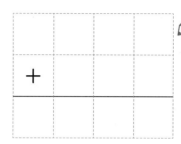

Way to see and think

Let's add the numbers in the same places.

Want to discuss

2 Find the mistakes in the following processes.

Let's talk about how to calculate correctly.

① 327 + 4   ② 649 + 13

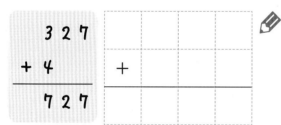
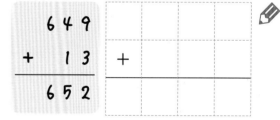

Want to confirm

3 Let's calculate the following in vertical form.

① 345 + 7   ② 286 + 4   ③ 121 + 9

④ 463 + 29   ⑤ 616 + 66   ⑥ 748 + 43

Want to solve Subtraction with borrowing from the hundreds place

1 We participated in cleaning the town. We picked up 73 cans last month and 129 cans this month. How many more cans did we pick up this month than last month?

① Let's write a math expression.

About how many more cans did you pick up?

Want to think

② Let's think about how to calculate.

Daiki's idea

129 can be decomposed into 100 and 29.
100 − 70 = 30
30 − 3 = 27
29 + 27 = 56

Yui's idea

129 can be decomposed into 120 and 9.
120 − 70 = 50
9 − 3 = 6
50 + 6 = 56

Want to communicate

③ Let's explain how to subtract 129 − 73 in vertical form.

	1	2	9
−		7	3

Subtraction algorithm for 129 − 73 in vertical form

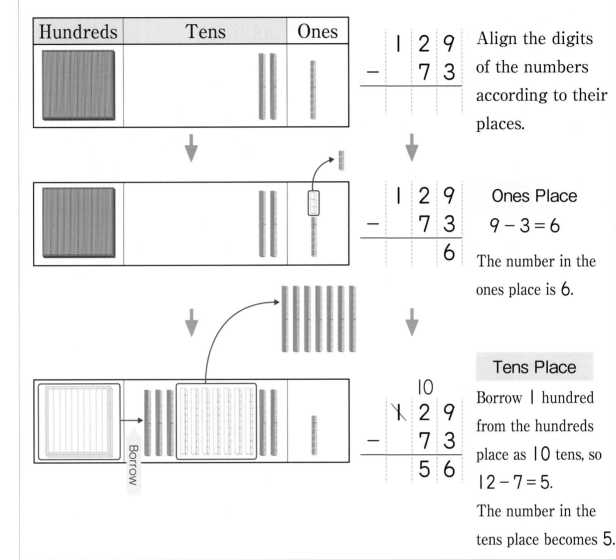

Hundreds	Tens	Ones

Align the digits of the numbers according to their places.

```
  1 2 9
−   7 3
```

Ones Place

$9 - 3 = 6$

The number in the ones place is 6.

```
  1 2 9
−   7 3
      6
```

Tens Place

Borrow 1 hundred from the hundreds place as 10 tens, so $12 - 7 = 5$.

The number in the tens place becomes 5.

```
    10
  ⨯ 2 9
−   7 3
    5 6
```

Math Sentence : 129 − 73 = 56   Answer : 56 cans

Want to confirm

1 ▶ Let's calculate the following in vertical form.

① 135 − 43   ② 154 − 92

③ 109 − 53   ④ 146 − 60

2 Let's explain how to subtract 125 − 86 in vertical form.

	1	2	5
−		8	6

Subtraction algorithm for 125 − 86 in vertical form

Hundreds	Tens	Ones

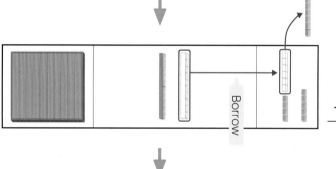

```
    1 2 5
  −   8 6
  ─────────
```

Align the digits of the numbers according to their places.

```
      1 10
    1 2̷ 5
  −   8 6
  ─────────
        ◯
```

Ones Place

Borrow 1 ten from the tens place as 10 ones, so 15 − 6 = 9.

The number in the ones place is 9.

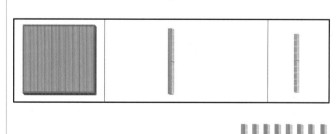

Tens Place

Borrow 1 hundred from the hundreds place as 10 tens, so 11 − 8 = 3.

The number in the tens place becomes 3.

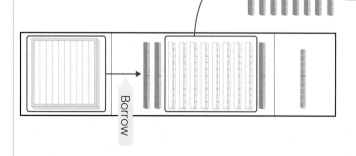

```
    10
     1 10
    1̷ 2̷ 5
  −   8 6
  ─────────
      ◯ 9
```

2 Let's calculate the following in vertical form.

① 156 − 78 ② 171 − 82 ③ 145 − 59 ④ 120 − 61

3 Let's think about how to subtract 105 − 78 in vertical form.

① Let's write in vertical form.

 Let's align the digits of the numbers.

② Let's explain how to subtract by using math expressions or diagrams.

Hundreds	Tens	Ones

Can I borrow 1 ten from the tens place?

What is the difference from the calculation of 125 − 86?

⊙ Purpose When 1 ten cannot be borrowed from the tens place, how can we subtract in vertical form?

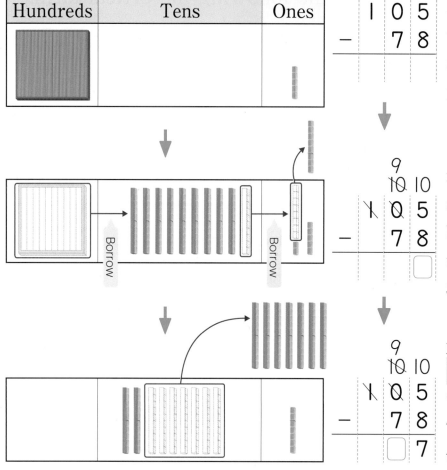

Align the digits of the numbers according to their places.

Ones Place

First borrow 1 hundred from the hundreds place as 10 tens, and then borrow 1 ten from the tens place as 10 ones. So 15 − 8 = 7.

The number in the ones place is 7.

Tens Place

9 − 7 = 2

The number in the tens place becomes 2.

🌸 Summary

When 1 ten cannot be borrowed from the tens place, borrow 1 hundred from the hundreds place to the tens place as 10 tens, and then borrow 1 ten from the tens place to the ones place as 10 ones.

Want to confirm

 Let's calculate the following in vertical form.

① 106 − 59   ② 103 − 44   ③ 101 − 83

④ 100 − 36   ⑤ 102 − 7   ⑥ 108 − 9

4 Subtraction of 3-digit numbers

1 Takaya and Marin would like to buy snacks that cost 300 yen.

Each of them has the money as shown below. How much of his or her money will be left after buying the snacks?

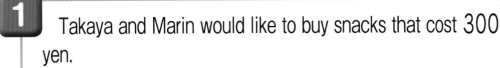

Way to see and think

Let's think about the number of 100-yen coins.

① Takaya has five 100-yen coins.

How much of his money will be left after buying the snacks?

Math Expression :

Answer :   yen

② Marin has ten 100-yen coins. How much of her money will be left after buying the snacks?

Math Expression :

Answer :   yen

 From where do we borrow?

Hiroto

1 Let's calculate the following.

① 900 − 500   ② 500 − 200   ③ 600 − 400

④ 700 − 100   ⑤ 800 − 300   ⑥ 1000 − 200

2 Let's calculate the following in vertical form.

① 753 − 6   ② 546 − 27

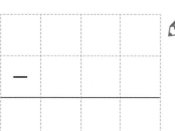   

Way to see and think

You have learned the calculation of 53 − 6 and 46 − 27.

Want to discuss

2 Find the mistakes in the following processes.

Let's talk about how to calculate correctly.

① 608 − 3   ② 524 − 17

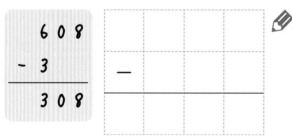   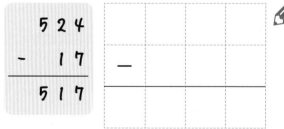

Want to confirm

3 Let's calculate the following in vertical form.

① 536 − 5   ② 273 − 4   ③ 115 − 8

④ 354 − 32   ⑤ 282 − 63   ⑥ 230 − 24

What you can do now

◻ Understanding the meaning of calculation and the way to calculate in vertical form.

1 Let's summarize how to calculate the following in vertical form.

① $73 + 56$

(1) In the ones place, $3 + 6 = 9$

(2) In the tens place, $7 + 5 = 12$

The tens place is ◻ . Carry ◻ hundred to the hundreds place.

(3) The answer is ◻ .

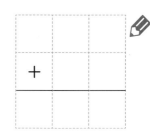

② $132 - 64$

(1) Borrow ◻ ten from the tens place, so the ones place will become ◻ $- 4 =$ ◻ .

(2) In the tens place, borrow ◻ hundred from the hundreds place, so ◻ $- 6 =$ ◻ .

(3) The answer is ◻ .

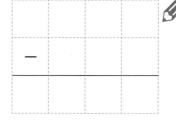

◻ Can calculate in vertical form.

2 Let's calculate the following in vertical form.

① $200 + 700$ ② $800 - 200$ ③ $1000 - 600$

④ $73 + 45$ ⑤ $30 + 80$ ⑥ $69 + 71$

⑦ $46 + 55$ ⑧ $8 + 95$ ⑨ $374 + 6$

⑩ $126 - 34$ ⑪ $148 - 67$ ⑫ $135 - 50$

⑬ $168 - 79$ ⑭ $107 - 48$ ⑮ $114 - 9$

◻ Can find the answers by making addition or subtraction expressions.

3 In Sota's school, there are 86 children in the 1st grade and 78 children in the 2nd grade. How many children are there in the 1st and 2nd grades altogether?

Supplementary Problems p. 139

Usefulness and efficiency of learning

1 Find the mistakes in the following processes.

Let's write the correct answer in the ( ).

Understanding the meaning of calculation and the way to calculate in vertical form.

①
```
    9 8
 +    9
 ─────
  1 1 7
```
( )

②
```
  1 2 1
 -  6 8
 ─────
    6 3
```
( )

③
```
  1 0 5
 -  1 7
 ─────
    9 8
```
( )

2 Let's calculate the following in vertical form.

Can calculate in vertical form.

① 300 + 500   ② 100 + 900   ③ 400 + 80

④ 600 − 100   ⑤ 700 − 500   ⑥ 1000 − 400

⑦ 38 + 67   ⑧ 35 + 65   ⑨ 436 + 8

⑩ 536 + 37   ⑪ 129 + 51   ⑫ 206 + 6

⑬ 106 − 43   ⑭ 130 − 51   ⑮ 374 − 7

⑯ 380 − 73   ⑰ 238 − 19   ⑱ 250 − 48

3 Yuri has 73 sheets of origami paper and her sister has 89 sheets.

Can find the answers by making addition or subtraction expressions.

① How many sheets of origami paper do they have altogether?

② Who has more sheets and by how many?

4 Let's make various addition expressions to make 1000 by using two numbers that are sets of 100.

Let's deepen.

I want to make problems by using addition or subtraction.

Yui

86

Deepen.

Let's make 100!

Want to think

Numbers from 1 to 9 are arranged in ascending order like '1 2 3 4 5 6 7 8 9.'

Let's make 100 that is the answer of a math expression by inserting + or − between numbers.

(example) $1 + 2 + 3 - 4 + 5 + 6 + 78 + 9 = 100$

① Let's fill in each ☐ with + or −.

If you insert neither + nor − between two numbers, the two numbers will be regarded as a 2-digit number.

1 ☐ 2 ☐ 3 ☐ 4 ☐ 5 ☐ 6 ☐ 7 ☐ 8 ☐ 9 = 100

1 ☐ 2 ☐ 3 ☐ 4 ☐ 5 ☐ 6 ☐ 7 ☐ 8 ☐ 9 = 100

1 ☐ 2 ☐ 3 ☐ 4 ☐ 5 ☐ 6 ☐ 7 ☐ 8 ☐ 9 = 100

1 ☐ 2 ☐ 3 ☐ 4 ☐ 5 ☐ 6 ☐ 7 ☐ 8 ☐ 9 = 100

Want to deepen

Numbers are arranged in descending order like '9 8 7 6 5 4 3 2 1.' Let's think about whether you can make 100.

Reflect Connect

Problem

Let's compare addition and subtraction in vertical form.

Let's calculate 59 + 73 in vertical form.

```
    5 9
  + 7 3
  ─────
  1 3 2
```

① Align the digits of the numbers according to their places.

② Ones Place

$9 + 3 = ①2$ — as 10 ones

Carry ① ten to the tens place.

③ Tens Place

$5 + 7 + 1 = ①3$ — as 10 tens

Carry 1 hundred to the hundreds place.

Add the numbers in the same places.

In fact it means '50 + 70 + 10 = ①30.'

Hundreds place

addition of 1-digit numbers

5+7+1=13
The answer is 13.
Why 1 hundred is carried to the hundreds place?

Daiki

In '5+7+1,' each number is in the tens place.
So it means '50+70+10.'

Let's subtract $146 - 78$ in vertical form.

```
      10
     3 10
   1̶ 4̶ 6
 -   7 8
 ─────────
     6 8
```

① **Align the digits of the numbers according to their places.**

② **Ones Place**
Borrow 1 ten from the tens place. ⌒ as 10 ones
$16 - 8 = 8$

③ **Tens Place**
Borrow 1 hundred from the hundreds place. ⌒ as 10 tens
$13 - 7 = 6$

> In fact it means
> '$130 - 70 = 60$.'

Summary : Between addition and subtraction in vertical form

Similarity	· To calculate the numbers in the same places. · To calculate in the ones place first.
Difference	· To carry in addition, to borrow in subtraction.

Let's look back

○ In vertical form, the calculation in each place is the same as the one we have learned in the 1st grade.
○ It seems that calculation of large numbers can be done as we have learned.

Calculation in the same places is addition or subtraction of 1-digit numbers that we have learned in the 1st grade.

Hiroto

By aligning the digits of the numbers according to their places, we can calculate the numbers in the same places.

Yui

I think we can find the answers of the calculation of much larger numbers by calculating in the same places.

Nanami

Which is longer?

How to Play

(1) Cut the tape on page 145.

(2) Let's play rock-paper-scissors game.

If you win with ✊, you get ▬ .
If you win with ✌, you get ▭ .
If you win with 🖐, you get ▭ .

Stick the tapes together once you win.

(3) One who gets the longest tape wins.

Let's play the rock-paper-scissors.

Rock, Paper, Scissors. One, Two, Three!

Which tape is longer?

Mine?

How about comparing them side by side?

It's not easy to move them.

Problem Let's think about how to compare lengths.

8 Length (1)

Let's think about how to compare lengths and how to represent it.

1 How to compare lengths

Want to compare

1 Nanami and Daiki played the rock-paper-scissors game. Let's compare the lengths of their tapes.

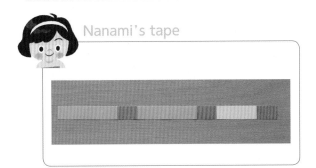
Nanami's tape

Daiki's tape

① Let's talk about how to compare the lengths.

Can you compare them as you have learned in the 1st grade?

How about aligning their edges?

But we have to move them.

We compared by counting the number of some units in the 1st grade.

Then we can compare the lengths without moving them.

Purpose How can we compare the lengths without moving them?

② How many ▬ did each one get?

Nanami

▬ ⬚ pieces

Daiki

▬ ⬚ pieces

③ Which is longer and by how many ▬ ?

🎯 Summary

   We can compare lengths by using the number of units of the same length.

Want to try

2 We put a book on a grid paper. Let's examine the length and width of the book.

① How many grids are the length and width of the book?

② Which is longer, the length or the width, and by how long?

▶1 Let's cut the grid paper to make a tool for measuring the length of various objects.

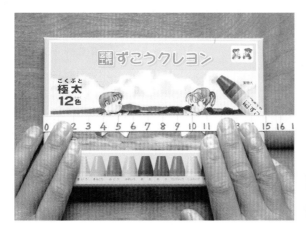

② How to represent lengths

Want to know Units of length

1 Let's measure the width of a postcard by using the scales on the grid paper.

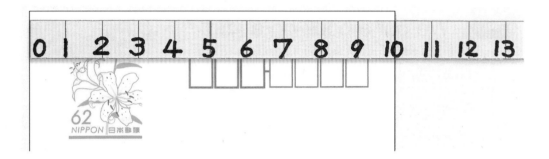

 There is a **unit** called **centimeter** to measure length. The length of one scale on the grid paper is written as 1 cm, and is read as " 1 centimeter."

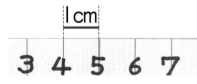

The width of a postcard is

[   ] cm.

Many countries use cm as a unit of length.

Want to confirm

 1 What is the length of the pencil in cm?

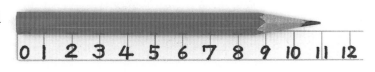

Which is the correct way to measure?

A 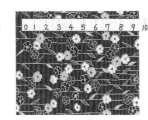 B 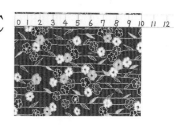 C

 Let's measure the lengths of the tapes and lines below.

① [gray tape]

② [gray tape] ☐ cm   ☐ cm

③ [black line] ☐ cm   ④ [line] ☐ cm

Want to find in our life

 Let's look for objects which are about 10 cm long.

It's a little more than 10 cm.

Nanami

Want to connect

How can we measure the length which is shorter than 1 cm?

Hiroto

2 How can you represent the length of this stick?

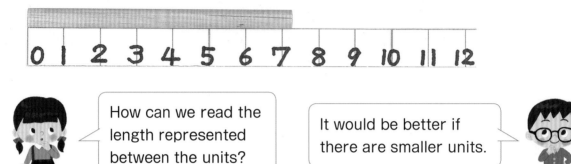

Yui: How can we read the length represented between the units?

Hiroto: It would be better if there are smaller units.

If you use a **ruler**, you can also measure the length which is shorter than 1 cm.

① The length of the stick is a little longer than 7 cm. How many smaller units are there beyond 7 cm?

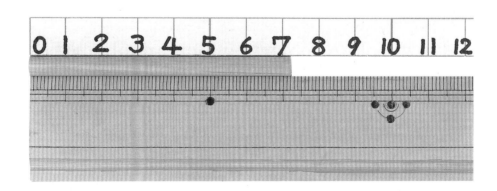

1 cm

② How many smaller units is 1 cm divided into?

When the length of 1 cm is equally divided into 10 parts, the unit of one part is written as 1 mm, and is read as "1 **millimeter**."

mm is also a unit used to measure length.

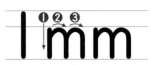

$$1\,cm = 10\,mm$$

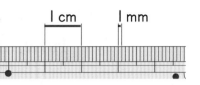

The length of the stick is 7 cm 2 mm. This is read as "7 centimeters and 2 millimeters."

Want to try

5 From the left end of the ruler, what are the lengths up to ①, ②, ③, and ④ in cm and mm?

Want to confirm

6 Let's measure the lengths of the tapes and lines.

① ⑤

②

③

④

Relationship between cm **and** mm

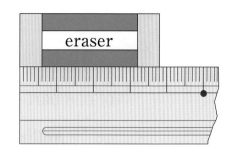

eraser

3 Let's measure the length of a side of an eraser.

① How many **cm** and **mm** is it?

② How many **mm** is it?

A. $3\,\text{cm} = \boxed{}\,\text{mm}$, so if we add $8\,\text{mm}$ to it, we will get $\boxed{}\,\text{mm}$.

$3\,\text{cm}\,8\,\text{mm} = \boxed{}\,\text{mm}$

cm	mm
3	8

B. There are 38 of $1\,\text{mm}$ units, so it will be $\boxed{}\,\text{mm}$.

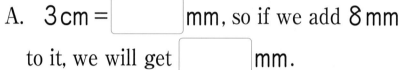

3 cm 8 mm

eraser

1 mm

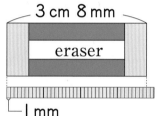

	mm
3	8

7 Let's fill in the $\boxed{}$ with numbers.

① $5\,\text{cm} = \boxed{}\,\text{mm}$ ② $7\,\text{cm}\,9\,\text{mm} = \boxed{}\,\text{mm}$

③ $80\,\text{mm} = \boxed{}\,\text{cm}$ ④ $62\,\text{mm} = \boxed{}\,\text{cm}\,\boxed{}\,\text{mm}$

How to draw a straight line

(1) Mark a point.

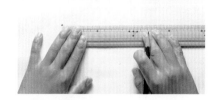

(2) Place one end of the ruler at the point.

(3) Mark another point that is $8\,\text{cm}$ from the first point.

4   Let's draw lines that don't curve for the given lengths with a ruler.

① 8 cm

② 11 cm 5 mm

③ 15 cm 8 mm

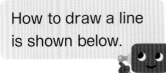
How to draw a line
is shown below.

A line that does not curve is called a **straight line**.

Want to try

8 ▶   Let's fill in the ☐ with >, <, or =.

And confirm by drawing the straight lines.

① 6 cm 3 mm ☐ 5 cm 8 mm

② 11 cm 5 mm ☐ 115 mm

③ 70 mm ☐ 7 cm 4 mm

01201

(4) Adjust the ruler to
connect the two points.

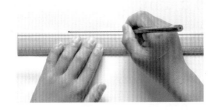

(5) Draw the line by
holding the ruler firmly.

Looking from the side.

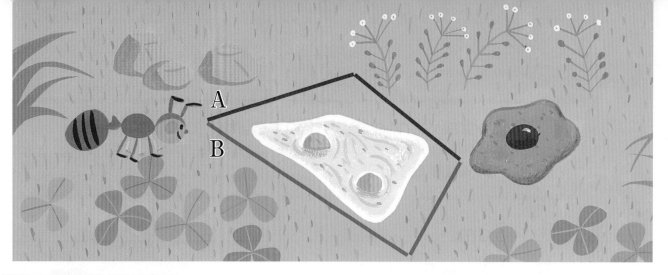

③ Calculating Length

Want to compare

1 Let's compare the length of the line A and the length of the line B.

Because the lines are not straight, we cannot measure them.
Daiki

Can we add or subtract lengths?
Nanami

Ⓨ **Purpose** Can we calculate lengths?

Want to think

① What is the length of the line A?

☐ cm ☐ mm + ☐ cm ☐ mm = ☐ cm ☐ mm

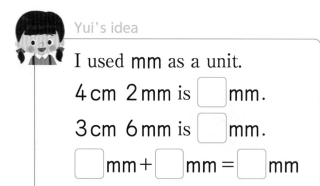
Yui's idea

I used mm as a unit.
4 cm 2 mm is ☐ mm.
3 cm 6 mm is ☐ mm.
☐ mm + ☐ mm = ☐ mm

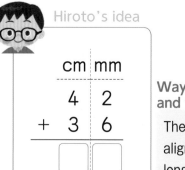
Hiroto's idea

```
    cm : mm
---------------
     4 :  2
+    3 :  6
---------------
   ☐ : ☐
```

Way to see and think
They both align the lengths with the same unit.

100

② What is the length of the line B?

$\boxed{}$ cm $\boxed{}$ mm + $\boxed{}$ cm $\boxed{}$ mm = $\boxed{}$ cm $\boxed{}$ mm

Want to try

③ What is the difference between the length of

the line A and the length of the line B?

$\boxed{}$ cm $\boxed{}$ mm − $\boxed{}$ cm $\boxed{}$ mm = $\boxed{}$ cm $\boxed{}$ mm

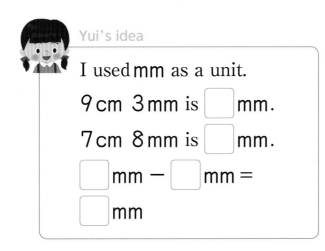

Yui's idea

I used mm as a unit.

9 cm 3 mm is $\boxed{}$ mm.

7 cm 8 mm is $\boxed{}$ mm.

$\boxed{}$ mm − $\boxed{}$ mm =

$\boxed{}$ mm

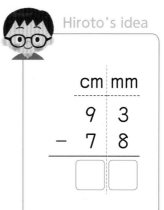

Hiroto's idea

cm	mm
9	3
− 7	8
$\boxed{}$	$\boxed{}$

Summary

To calculate lengths, we should add or subtract numbers in the same units.

Want to confirm

 Let's calculate the following.

① 12 cm + 25 cm ② 13 cm 5 mm + 4 cm 7 mm

③ 28 cm − 16 cm ④ 23 cm 6 mm − 2 cm 9 mm

What you can do now

☐ Can measure the length with a ruler.

1 From the left end of the ruler, what are the lengths up to ①, ②, ③, and ④ in cm and mm?

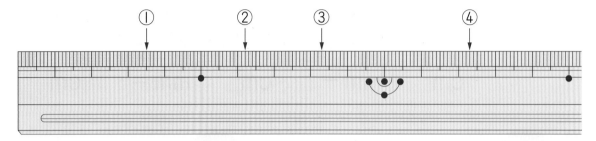

☐ Understanding the conversion of the units of lengths.

2 Let's fill in the ☐ with numbers.

① 4 cm = ☐ mm　　② 2 cm 5 mm = ☐ mm

③ 8 cm 7 mm = ☐ mm　　④ 60 mm = ☐ cm

⑤ 23 mm = ☐ cm ☐ mm

☐ Can compare lengths.

3 Let's fill in the ☐ with >, <, or =.

① 7 cm 2 mm ☐ 6 cm 9 mm

② 12 cm 3 mm ☐ 123 mm

☐ Can calculate length.

4 Let's calculate the following.

① 6 cm 5 mm + 5 cm

② 3 cm 4 mm + 2 cm 3 mm

③ 23 cm 8 mm − 8 cm

④ 8 cm 9 mm − 6 cm 5 mm

Supplementary Problems •••••••• ➤ p. 140

Usefulness and efficiency of learning

1 Let's measure the lengths of the following tapes and straight lines.

①
②
③

☐ Can measure the length with a ruler.

2 Let's draw straight lines for the given lengths in your notebook.

① 5 cm

② 8 cm 3 mm

③ 10 cm 6 mm

☐ Can measure the length with a ruler.

3 Let's calculate the following and draw straight lines for the lengths of the answers with a ruler.

① 10 cm 5 mm + 8 cm 7 mm

② 15 cm 2 mm − 6 cm 7 mm

☐ Can calculate length.

4 Let's arrange the following lengths from the longest to the shortest.

71 mm   6 cm 9 mm   7 cm

☐ Can compare lengths.

5 There are two tapes.

① Let's find the sum of their lengths.

② Let's find the difference of their lengths.

☐ Can calculate length.

To what should we pay attention when we subtract in vertical form?

Want to explore Subtraction in vertical form

They solved the problems of subtraction in vertical form in their mathematics class.

Then **6** mistakes were found in the algorithms.

①

$$\begin{array}{r} 138 \\ -76 \\ \hline 162 \end{array} \qquad \begin{array}{r} 159 \\ -89 \\ \hline 170 \end{array}$$

②

$$\begin{array}{r} 322 \\ -14 \\ \hline 312 \end{array} \qquad \begin{array}{r} 135 \\ -73 \\ \hline 142 \end{array}$$

③

$$\begin{array}{r} 141 \\ -69 \\ \hline 82 \end{array} \qquad \begin{array}{r} 105 \\ -68 \\ \hline 47 \end{array}$$

Each algorithm has borrowing.

Nanami

1 Let's think about how mistakes are made in the algorithm ①.

Yui

In the tens place, they borrowed ....

Let's compare them to the correct answers.

Hiroto

2 Make two groups. Let's talk about how mistakes are made in the algorithm ② in one group and in the algorithm ③ in the other.

If you find how to make mistakes in one algorithm, try to find in the other.

3 In your group, let's summarize the mistakes in the way to subtract in vertical form and what you should pay attention to when you subtract in vertical form.

Then let's confirm them in the class.

Let's summarize what you should pay attention to when you also add in vertical form.

01202

Which is more?

Problem Let's compare which bottle can hold more water.

9 Amount of Water

Let's think about how to compare the amounts of water and how to represent it.

❶ How to compare the amounts of water

Want to compare

1 Yui's water bottle contains 5  of water and Hiroto's contains 6 of water.

Let's compare the amounts of water that each bottle can hold.

① Can we say which one holds more?

The one that can be filled by 6 cups may contain more.

But the sizes of the cups ....

Nanami

Daiki

A quantity of something like water is called an amount.

♈ Purpose How can we compare the amounts?

② Let's try by using the same cup. Which one contains more?

Yui

 ☐ cups

Hiroto

 ☐ cups

❀ Summary

We can compare amounts by the number of units of amount used.

2 How to represent the amount of water

We can compare the amounts of water by counting the number of filled **unit cup**s.

There is a unit called **liter** to represent the amount of water.

I litter is written as I L.

Each container in the picture above can hold I L.

In many countries, liter is used as a unit to measure the amount of liquid.

1 The following amount of water was measured by using I L measuring cups. How many liters is each amount of water?

① Plastic bottle

② Bucket

[   ] L

[   ] L

 Let's look for containers with L as the unit of measure.

2 The amount of water in the pot was measured by using 1 L measuring cups. How can we measure the portion that is less than 1 L?

To measure the portion that is less than 1 L, we can use a 1 **deciliter** measuring cup.

The amount of water in the pot was 2 L and 3 deciliter measuring cups. This is read as "two liters and three deciliters."

 Let's fill a 1 L measuring cup with water by using a 1 deciliter measuring cup.

How many times do we need to use the cup?

When 1 L is divided into 10 equal amounts, the unit of 1 amount is called 1 deciliter.

1 deciliter is written as 1 dL.

Deciliter is another unit to represent the amount of liquid.

$$1 L = 10 dL$$

Want to explore Measuring the amount

3   Let's measure the amounts of water that can be placed into the following containers.

①   ☐ L ☐ dL

②   ☐ L ☐ dL

③   ☐ dL

3 Let's estimate the amounts of water in various containers and measure them.

4 Water in the plastic bottle was poured into the | L measuring cup as shown by the figure below.

How many deciliters of water were there in the plastic bottle?

One scale represents one part of the | L that has been divided into |0 ....

Yui

4 Let's measure the amount of water that can be placed into the following container.

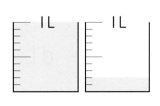

That's it! 💡 **Making a 1 dL Measuring Cup**

Fill the |dL cup with water, pour the water into a container, and put a mark on the container to indicate where the water level is.

Relationship between L and dL

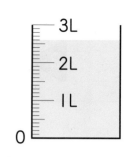

3L
2L
1L
0

5 Let's find the amount of water in the container on the right.

① How many L and dL are there?

2 L and 6 smaller scales is

☐ L ☐ dL.

L	dL
2	6

② How many dL are there?

2 L = ☐ dL, so when it is added to 6 dL, the total is ☐ dL.

	dL
2	6

5 How many L and dL are there in 54 dL?

6 Let's fill in the ☐ with numbers.

① 3 L = ☐ dL   ② 4 L 9 dL = ☐ dL

③ 68 dL = ☐ L ☐ dL

7 Let's fill in the ☐ with >, <, or =.

① 2 L 5 dL ☐ 1 L 5 dL   ② 1 L 3 dL ☐ 13 dL

③ 70 dL ☐ 7 L 2 dL

6 Let's use a 1 dL measuring cup to find the amount of juice in the can.

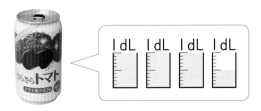

To represent the amount of water, there is a unit called **milliliter** that is smaller than L and dL.

1 milliliter is written as 1 mL.

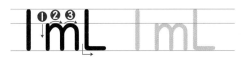

Want to confirm

8 Let's use a 1 L measuring cup and 1 dL measuring cups to find the amount of water in a 1000 mL pack.

We can write liter as L, l, or ℓ. Milliliter can also be written as mL, ml, or mℓ. And 1 mL can be called 1 cc.

| 1 L = 1000 mL | | 1 dL = 100 mL |

Want to find in our life

9 Let's look for containers with mL as the unit of measure.

200ml
要冷蔵10℃以下

1日分の
カルシウム
種類別
乳飲料

200ml

1 There is 2 L 5dL of water in a basin and
I L 3dL of water in a flower vase.

① How many L and dL are there altogether?

Math Expression :

② Let's think about how to calculate.

Way to see
and think

They both align
the lengths
with the same
unit.

Hiroto's idea

I came up with
the two amounts of
water in dL.

2 L 5dL is ☐ dL.

I L 3dL is ☐ dL.

Yui's idea

I aligned the
numbers with the
same unit in the same
column and then
added the numbers in
each column.

L	dL
2	5
+ 1	3
☐	☐

③ What is the difference between the amounts
of water in the basin and in the flower vase?

Way to see
and think

Can we use
the same idea
as in addition?

1 Let's calculate the following.

① 7 L + 2 L   ② 3 L 2dL + I L 5dL

③ 3 L 6dL + I L 8dL   ④ 500 mL − 200 mL

⑤ 6 L 4dL − I L 3dL   ⑥ 7 L − 3 L 5dL

What you can do now

☐ Can represent the amount of water.

1 How much is the amount of water?

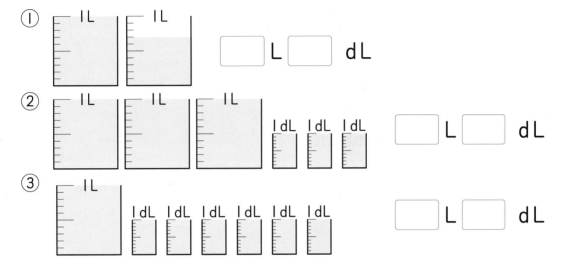

① [　] L [　] dL

② [　] L [　] dL

③ [　] L [　] dL

☐ Understanding the conversion of the units of amount.

2 Let's fill in the [　] with numbers.

① 1 L = [　] dL   ② 1 L = [　] mL

③ 1 dL = [　] mL

☐ Can compare the amounts of water.

3 Let's fill in the [　] with >, <, or =.

① 9 dL [　] 8 dL   ② 2 L [　] 20 dL

③ 450 mL [　] 5 dL   ④ 3 L 2 dL [　] 31 dL

☐ Can calculate the amount of water.

4 Let's calculate the following.

① 1 L 3 dL + 7 dL   ② 2 L + 1 L 5 dL

③ 4 L 5 dL + 1 L 6 dL   ④ 3 dL + 2 L 8 dL

⑤ 3 L 6 dL − 1 L 5 dL   ⑥ 2 L 4 dL − 4 dL

⑦ 6 L 4 dL − 2 L 6 dL   ⑧ 1 L − 7 dL

Supplementary Problems p. 141

Usefulness and efficiency of learning

1 Let's fill in the ☐ with numbers.

① [measuring cups illustration]

☐ L ☐ dL

② [measuring cup illustration]

☐ dL

Can represent the amount of water.

2 Let's fill in the ☐ with numbers.

① $5 L = $ ☐ dL

② $75 dL = $ ☐ L ☐ dL

③ $4 dL = $ ☐ mL

④ $3 L 8 dL = $ ☐ dL

⑤ $900 mL = $ ☐ dL

Understanding the conversion of the units of amount.

3 Let's arrange the amounts of liquid below from the most to the least.

$36 dL$ $360 mL$ $36 L$

Can compare the amounts of water.

4 There are two containers that have water.

 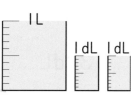  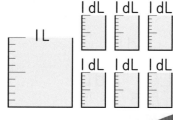

Can calculate the amount of water.

① How many L and dL of water are there altogether?

② What is the difference between the amounts of water in the 2 containers?

Let's deepen.

Let's think about how to measure the amount of water without using a 1 L measuring cup or a 1 dL measuring cup.

Daiki

116

Deepen.

Let's make 6 dL!

Want to know

There are 5 dL and 8 dL cups as well as a big cup.

How do you make 6 dL by using these cups?

A    B 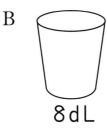   C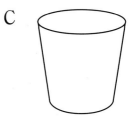

  5 dL   8 dL

Want to represent

① Yui has the following idea. Let's continue writing her idea.

Yui's idea

If I put two cups of B into cup C, then there will be ☐ dL of water in cup C.

Then,

Want to deepen

Let's try to think of other ways.

10 Triangle and Quadrilateral
Let's examine the shapes and sort them.

1 Triangle and quadrilateral

Want to try

1 Let's connect the points by using straight lines to enclose the animals.

Try to enclose each animal with fewest lines.

Want to explore

1 Let's sort the shapes that are enclosed by straight lines into **2** groups.

Way to see and think
What is the similarity and difference between two groups?

I sorted them by the number of straight lines.

Daiki

B

A

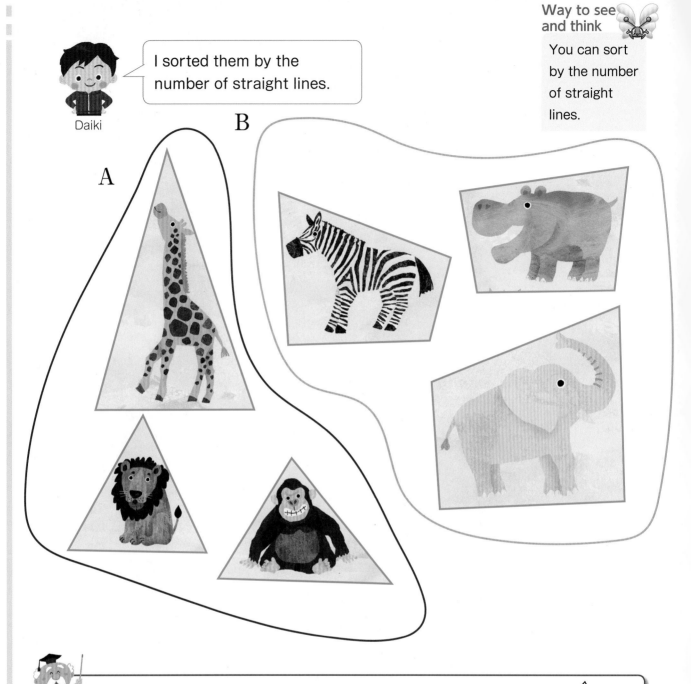

The shape enclosed by 3 straight lines is called a **triangle**.

The shape enclosed by 4 straight lines is called a **quadrilateral** (or **quadrangle**).

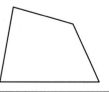

2 Let's find triangles and quadrilaterals.

Way to see and think

Let's think about why some of the shapes are neither triangles nor quadrilaterals.

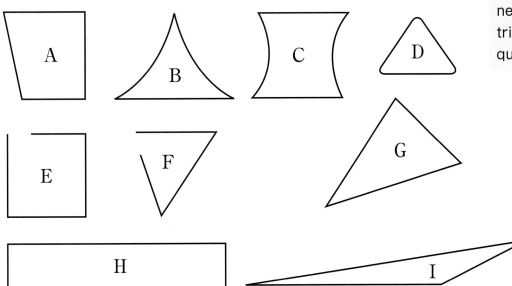

 Want to try

2 ▶ Let's find triangles and quadrilaterals and trace them.

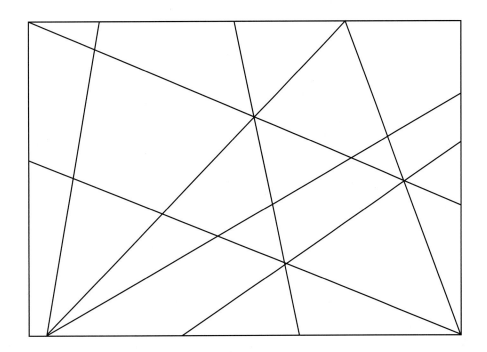

3 Let's draw various triangles and quadrilaterals by connecting points with straight lines.

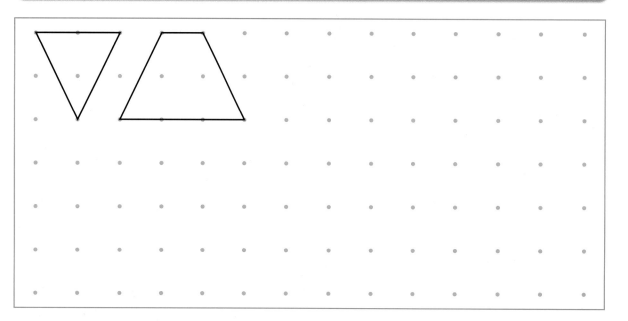

 Each straight line around triangles and quadrilaterals is called a **side** and each corner point that two sides make is called a **vertex**.

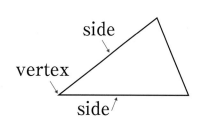

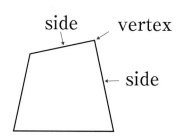

In a triangle, the number of sides is ☐ and the number of vertices is ☐.

In a quadrilateral, the number of sides is ☐ and the number of vertices is ☐.

4 Let's draw a straight line in the quadrilateral and make the following shapes.

① **2** triangles

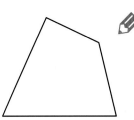

② **1** triangle and **1** quadrilateral

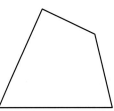

③ **2** quadrilaterals

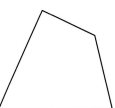

Let's explain how to draw them.

3 Let's draw a straight line in the triangle and make the following shapes.

① **2** triangles

② **1** triangle and **1** quadrilateral

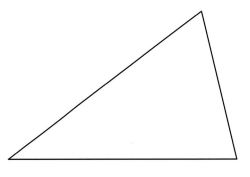

2 Right angle

1 Let's fold a sheet of paper as shown below.

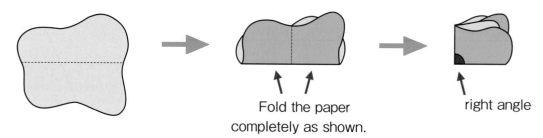

Fold the paper
completely as shown.

right angle

The corner that is formed by folding the paper as shown above is called a **right angle**.

1 Is there a right angle in a triangle ruler? Let's check and see.

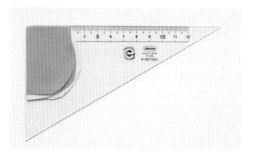

2 From your surroundings, let's look for right angles.

It seems that there are also right angles found in your school.

Nanami

2 Let's draw many shapes with right angles by connecting the dots below.

Let's check the right angles by using a triangle ruler.

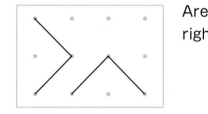

Are these right angles?

Daiki

3 Let's draw a right angle using a triangle ruler.

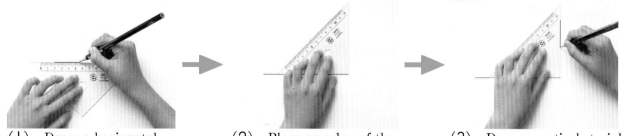

(1) Draw a horizontal straight line.

(2) Place an edge of the triangle ruler along the line you have just drawn.

(3) Draw a vertical straight line.

01203

❸ Rectangle and square

  Rectangle

1 Let's fold a sheet of paper as shown below and make right angles.

What shape is formed?

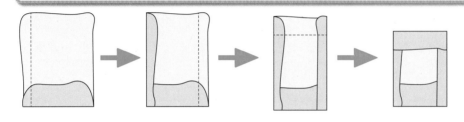

The third right angle is formed at the same time as the fourth.

 A quadrilateral where all 4 corners are right angles is called a **rectangle**.

Want to confirm

 1 Which ones are rectangles?

A

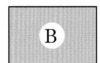
B

C

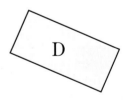
D

Want to find in our life

 2 From your surroundings, let's look for things shaped rectangles.

2 Let's compare the lengths of the opposite sides of a rectangle.

Measure and compare.

Fold and compare.

The lengths of the opposite sides of a rectangle are the same.

Be sure that there are no overlaps.

3 Let's draw the following rectangles.

① The lengths of the sides are 3 cm and 6 cm.

② The lengths of the sides are 1 cm and 7 cm.

③ The lengths of the sides are 5 cm and 4 cm.

1cm

1cm

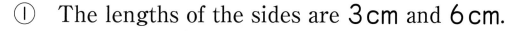

3 Cut the rectangular paper as shown below. Open the folded paper. What shape is formed?

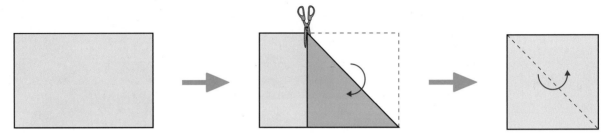

① Let's examine the four corners of the shape.

② Let's examine the four sides of the shape.

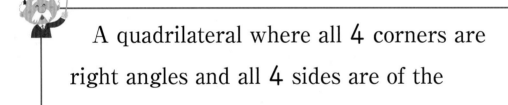

A quadrilateral where all 4 corners are right angles and all 4 sides are of the same length is called a **square**.

4 Which ones are squares?

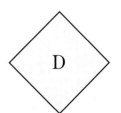

A   B   C   D

5 From your surroundings, let's look for things shaped squares.

4 Right triangle

1 Let's cut rectangular and square sheets of paper along the dashed lines as shown in the diagram. Look at the shapes formed by cutting.

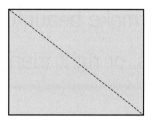 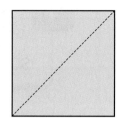

① Let's examine the shapes of the corners.

② What shapes are formed?

A triangle that has a right angle corner is called a **right triangle**.

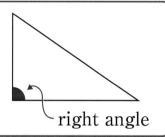

right angle

1 Which ones are right triangles?
Let's confirm by using a triangle ruler.

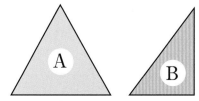 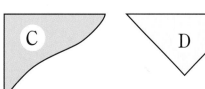

A B C D

Can you guess before using a triangle ruler?

1 Let's make beautiful patterns by drawing rectangles, squares, or right triangles using the dots below.

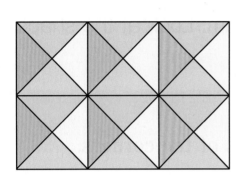

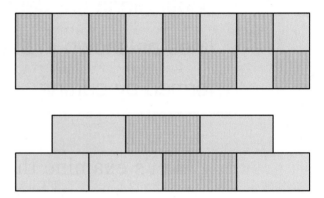

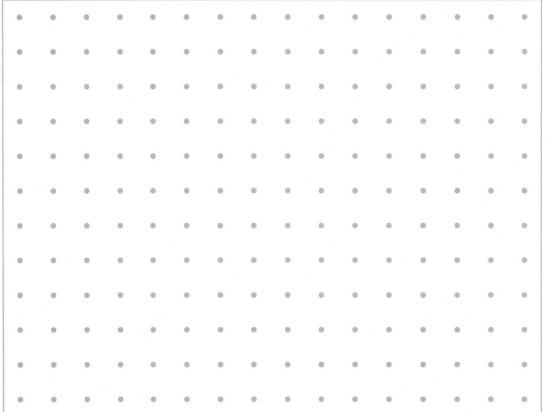

What you can do now

☐ Understanding the properties of triangles and quadrilaterals.

1 Let's fill in each ☐ with a number.

① There are ☐ sides and ☐ vertices in a triangle.

② There are ☐ sides and ☐ vertices in a quadrilateral.

☐ Understanding the properties of rectangles, squares, and right triangles.

2 Which of the following are rectangles, squares, and right triangles?

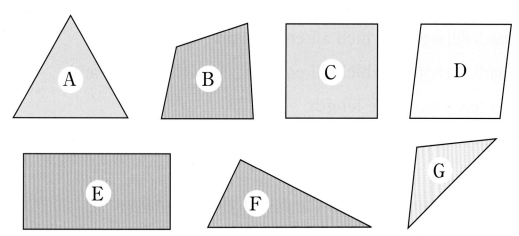

☐ Can write a square and a right triangle.

3 Let's draw the following shapes.

① A right triangle

② A square with a side of 3 cm

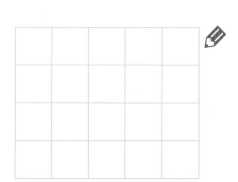

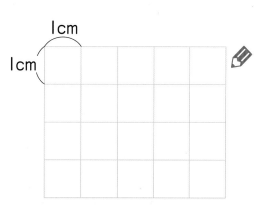

Usefulness and efficiency of learning

1 Which of the following are triangles and quadrilaterals?

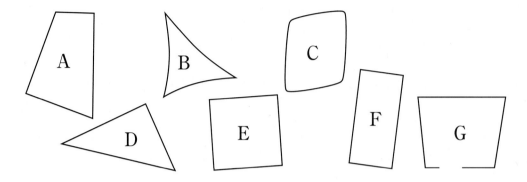

2 Identify the shape.

① A quadrilateral in which all corners are right angles.

② A quadrilateral in which all corners are right angles and all sides have the same length.

③ A triangle with a right angle.

3 Let's draw a straight line in the rectangle and make the following shapes.

① **2** right triangles

② **2** squares

Supplementary Problems

1 Tables and Graphs
pp.11~17

1   The weather in March is shown in the table below.

Let's answer the following.

Weather in March

Weather	Sunny	Cloudy	Rainy	Snowy
Number of days	12	9	8	2

①   Represent the number of days in each weather condition by using ○ on the graph.

Weather in March

Sunny	Cloudy	Rainy	Snowy

②   How many more sunny days are there than cloudy days?

2 Time and Duration(1)
pp.18~26

1   Let's answer the following.

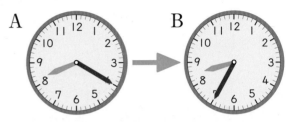

①   Express the time shown on clocks A and B.

②   How long is it from the time A to the time B?

2   How long is it from the time C to the time D as shown below?

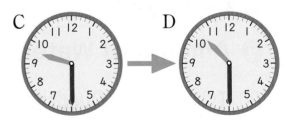

3 Let's fill in each ▢ with a number.

① 1 hour = ▢ minutes

② The short hand of the clock rotates completely around the clock ▢ times a day.

The first complete rotation of the short hand makes ▢ hours in the morning.

The next complete rotation of the short hand makes ▢ hours in the afternoon.

③ 1 day = ▢ hours

4 Addition in Vertical Form
pp.35～45

1 Let's calculate the following in vertical form.

① 23 + 34   ② 65 + 12

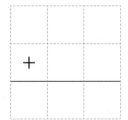

③ 29 + 40   ④ 70 + 18

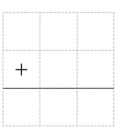

2 Let's calculate the following in vertical form.

① 42 + 36   ② 26 + 33
③ 75 + 21   ④ 36 + 20
⑤ 40 + 41   ⑥ 10 + 80

3 Let's calculate the following in vertical form.

① 5 + 43   ② 8 + 31
③ 74 + 4   ④ 52 + 7

4 There are 23 sheets of red paper and 5 sheets of blue paper.

How many sheets are there altogether?

5 Let's calculate the following in vertical form.

① 18 + 35   ② 46 + 26

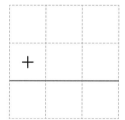

③ 39 + 42   ④ 58 + 37

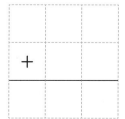

6 Let's calculate the following in vertical form.

① 18 + 34   ② 47 + 35
③ 55 + 29   ④ 38 + 13

7 Let's calculate the following in vertical form.

① 38 + 32   ② 74 + 6
③ 19 + 41   ④ 63 + 9
⑤ 4 + 78   ⑥ 3 + 57

8 Let's find the mistakes in the following processes and correct them.

①
```
    3 6
  + 2 8
  -----
    5 4
```

②
```
    5 9
  +   4
  -----
    9 9
```

③
```
    3 3
  +   5
  -----
    8 8
```

④
```
    6 3
  +   7
  -----
    6 0
```

9 The answer of the addition sentence below is 70.

Let's fill in the ☐ with a number.

20 + ☐ = 70

---

⑤ **Subtraction in Vertical Form**

pp.46~56

1 Let's calculate the following in vertical form.

① 43 − 21   ② 68 − 56

 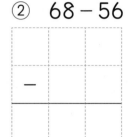

③ 37 − 31   ④ 59 − 8

 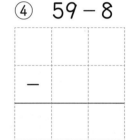

2 Let's calculate the following in vertical form.

① 87 − 52   ② 65 − 20
③ 55 − 31   ④ 94 − 83

3 There were 64 sheets of origami paper. I used 33 sheets.

How many sheets are left?

4 Let's calculate the following in vertical form.

① 49 − 3   ② 56 − 26
③ 38 − 31   ④ 74 − 4

5 Let's calculate the following in vertical form.

① 34 − 16   ② 60 − 23

③ 52 − 47   ④ 70 − 65

6 Let's calculate the following in vertical form.

① 53 − 14   ② 98 − 59
③ 30 − 19   ④ 60 − 47
⑤ 44 − 36   ⑥ 72 − 63

7 There are 35 children in Taiga's class. 17 of them are boys. How many are girls?

8 Let's calculate the following in vertical form.

① 34 − 5   ② 52 − 9
③ 80 − 7   ④ 30 − 6

9 There were 25 cookies. You ate 9 cookies. How many cookies are left?

10 Let's find the mistakes in the following processes and correct them.

①   6 3
     − 2 9
     ─────
        4 4

②   8 5
     −   4
     ─────
        4 5

③   7 8
     − 3 5
     ─────
        3 3

④   6 0
     −   7
     ─────
        6 3

11 Let's calculate the following in vertical form. And confirm the answer.

① 48 − 16   ② 52 − 29
③ 27 − 9   ④ 90 − 8

12 The answer of the subtraction sentence below is 20.

Let's fill in the ☐ with a number.

50 − ☐ = 20

6 Numbers up to 1000

pp.57～70

1 Let's write the following numbers.

① four hundred seventy-five

② six hundred twenty-eight

③ The number when five hundred, thirty, and seven are added.

④ The number when eight hundred, ninety, and two are added.

⑤ The sum of 6 sets of 100, 5 sets of 10, and 4 ones.

⑥ The sum of 3 sets of 100, 1 set of 10, and 7 ones.

2 Let's write the following numbers.

① nine hundred eighty

② five hundred three

③ The number that has 6 in the hundreds place, 0 in the tens place, and 2 in the ones place.

④ The number that is the sum of 7 sets of 100.

3 Let's fill in each ☐ with a number.

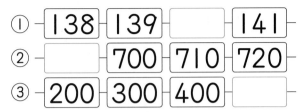

① 138 - 139 - ☐ - 141

② ☐ - 700 - 710 - 720

③ 200 - 300 - 400 - ☐

4 Let's write the following numbers.

① The number that is 300 larger than 400.

② The number that is 200 smaller than 800.

③ The number that is 100 smaller than 1000.

④ The number that is 10 larger than 600.

⑤ The number that is 10 smaller than 900.

5 Let's write the numbers that each ↑ is pointing at.

①

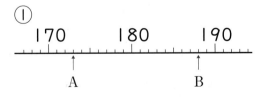

②

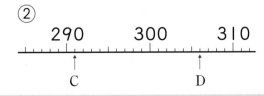

6 Let's fill in each ☐ with a number.

① 270 is the sum of ☐ sets of 100 and ☐ sets of 10.

② 270 is the sum of ☐ sets of 10.

③ 600 is the sum of ☐ sets of 10. Also, it is the sum of ☐ sets of 100.

④ The number that is the sum of 7 sets of 100 and 3 sets of 10 is ☐.

⑤ The number that is the sum of 45 sets of 10 is ☐.

⑥ The number that is the sum of 90 sets of 10 is ☐.

7 Let's fill in the ☐ with > or <.

① 685 ☐ 593
② 776 ☐ 767
③ 394 ☐ 396
④ 401 ☐ 410

7 Addition and Subtraction of Large Numbers

pp.71～87

1 Let's calculate the following in vertical form.

① 54 + 63   ② 46 + 80

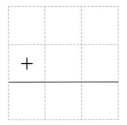 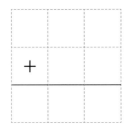

2 Let's calculate the following in vertical form.

① 63 + 75   ② 85 + 41
③ 52 + 92   ④ 69 + 60
⑤ 70 + 88   ⑥ 90 + 30

3 Let's calculate the following in vertical form.

① 46 + 87   ② 77 + 55
③ 95 + 26   ④ 62 + 68
⑤ 63 + 39   ⑥ 87 + 16
⑦ 45 + 55   ⑧ 8 + 92

4 Let's calculate the following in vertical form.

① 300 + 300   ② 500 + 200
③ 400 + 70   ④ 600 + 50
⑤ 500 + 500   ⑥ 300 + 700

5 Let's calculate the following in vertical form.
   ① 285 + 8   ② 769 + 1
   ③ 173 + 7   ④ 145 + 26
   ⑤ 328 + 54   ⑥ 846 + 35

6 Let's calculate the following in vertical form.
   ① 127 − 45   ② 168 − 83

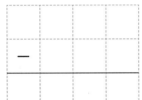 

7 Let's calculate the following in vertical form.
   ① 146 − 53   ② 159 − 85
   ③ 119 − 72   ④ 163 − 93
   ⑤ 104 − 11   ⑥ 125 − 60

8 Let's calculate the following in vertical form.
   ① 147 − 68   ② 173 − 95
   ③ 182 − 86   ④ 151 − 73
   ⑤ 124 − 57   ⑥ 160 − 91

9 Let's calculate the following in vertical form.
   ① 104 − 38   ② 106 − 19
   ③ 100 − 84   ④ 107 − 8

10 Let's calculate the following in vertical form.
   ① 800 − 400   ② 500 − 100
   ③ 900 − 300   ④ 700 − 200
   ⑤ 1000 − 500

11 Let's calculate the following in vertical form.
   ① 426 − 4   ② 632 − 3
   ③ 114 − 8   ④ 578 − 67
   ⑤ 345 − 36   ⑥ 260 − 53

12 Let's find the mistakes in the following processes and correct them.

①
```
   1 0 6
 −   4 8
 ───────
     6 8
```

②
```
   4 5 3
 −   2 6
 ───────
   1 9 3
```

8 Length (1)
pp.90~103

1 What is the longest tape?

A
B
C

2 What are the lengths of the tapes and the straight lines below in cm?

①
②
③
④

3 What are the lengths of the tapes and the straight line below in cm and mm?

①
②
③

4 Let's draw straight lines for the given lengths.

① 9 cm
② 10 cm 5 mm
③ 8 cm 7 mm

5 Let's fill in each ☐ with a number.

① 7 cm = ☐ mm
② 2 cm 8 mm = ☐ mm

③ 80 mm = ☐ cm
④ 93 mm = ☐ cm ☐ mm

6 Let's fill in the ☐ with >, <, or =.

① 4 cm 8 mm ☐ 52 mm
② 9 cm ☐ 89 mm
③ 106 mm ☐ 10 cm 6 mm

7 Let's calculate the following.

① 42 cm + 36 cm
② 15 cm 4 mm + 9 cm
③ 33 cm − 18 cm
④ 17 cm 5 mm − 9 cm

9 Amount of Water
pp.106〜117

1 How many L and dL of water are there?

①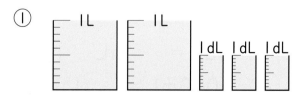

②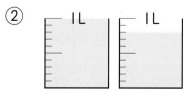

2 Let's fill in each ☐ with a number.

① 6 L = ☐ dL

② 3 L 7 dL = ☐ dL

③ 90 dL = ☐ L

④ 42 dL = ☐ L ☐ dL

3 Let's fill in the ☐ with >, <, or =.

① 3 L 4 dL ☐ 34 dL

② 1 L 8 dL ☐ 20 dL

③ 6 L 3 dL ☐ 62 dL

4 Let's calculate the following.

① 2 L 6 dL + 4 dL

② 4 L 8 dL + 2 L 7 dL

③ 8 L − 6 L 8 dL

④ 5 L 7 dL − 2 L 5 dL

5 Let's fill in each ☐ with a number.

① 3 dL = ☐ mL

② 1000 mL = ☐ L

③ 600 mL = ☐ dL

⑩ **Triangle and Quadrilateral**
pp.118~132

1 Let's fill in each ☐ with a word or a number.

①
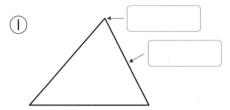

In a triangle, the number of sides is ☐ and the number of vertices is ☐ .

②
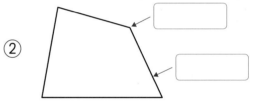

In a quadrilateral, the number of sides is ☐ and the number of vertices is ☐ .

2 Which corners are right angles?

①
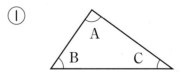

②
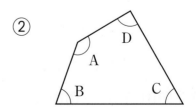

3 Which one is a rectangle?

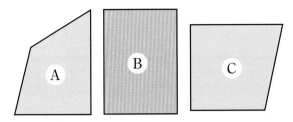

4 Look at the following rectangle and let's fill in each ☐ with a number.

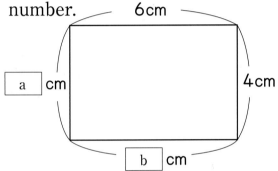

5 Which one is a square?

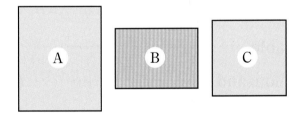

6 Which one is a right triangle?

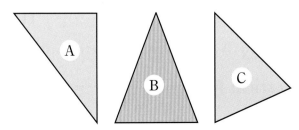

7 Let's draw the following shapes.

① A rectangle with sides of 3cm and 5cm.

② A square with a side of 4cm.

③ A right triangle.

Symbols and Words in This Book

Tape

▼ will be used in pages 90 and 91.

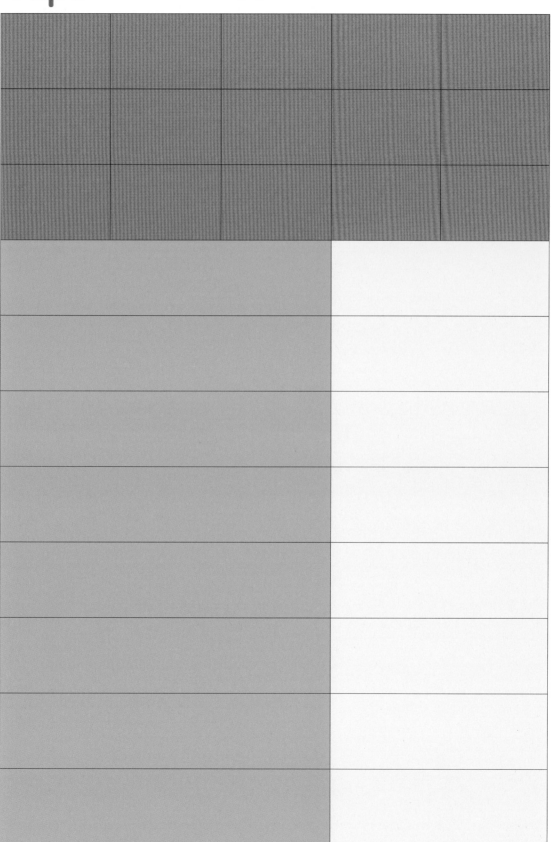

Memo